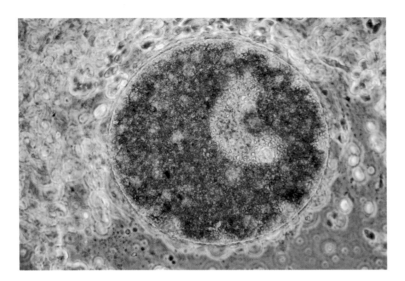

Frontispiece. (a) *Ichthyophthirius multifiliis*. The adult stage of this parasite has a characteristic horseshoe-shaped nucleus. It is usually brown-coloured and revolves slowly. It is approximately 0.5 mm in size and may be seen in skin scrapings under low power magnification.

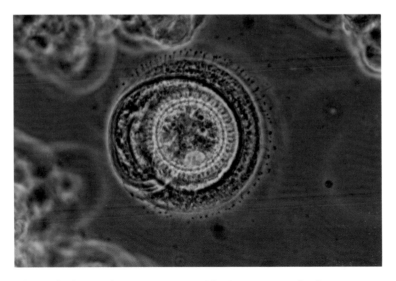

(b) *Trichodina*. This parasite, with its saw-toothed structure, revolves very slowly and is readily seen under low power in gill and skin smears.

(Photographs (a) and (b) courtesy of C. H. Aldridge.
Copyright, Unilever Research Laboratories.)

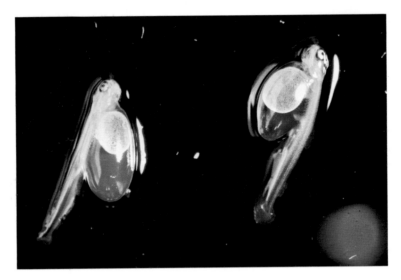

(c) Blue-sac disease in rainbow trout sac fry.
(Courtesy of Dr T. Håstein.)

(d) Furunculosis in a brown trout. This specimen shows the characteristic raised furuncles. The fish was taken in autumn and hence the rather uncharacteristic concurrent fungus infection. (Courtesy of G. Macgregor.)

Handbook of Trout and Salmon Diseases

Third Edition

The Authors

Emeritus Professor Ronald J Roberts worked for seven years as a lecturer in the University of Glasgow Veterinary School prior to establishing the Institute of Aquaculture, at the University of Stirling, where he was Director for 25 years before retiring in 1996.

He is best known for his research on skin diseases of fish and for his pioneering work on tropical fish pathology development and training in South East Asia.

He has been the recipient of international prizes and medals for his work and in 1992 he was invested as a Commander of the Order of the Crown, by HM The King of Thailand, only the second foreign citizen to be so honoured.

He is a non-executive director of Landcatch Ltd, the leading international salmon egg and smolt producer and is editor of *Aquaculture Research* and *Journal of Fish Diseases*.

Jonathan Shepherd has been in the aquaculture industry for 25 years since qualifying as a veterinary surgeon. After a period as Deputy Director of Stirling University's Institute of Aquaculture, he embarked on a commercial career, which has included general management posts in Unilever and Norsk Hydro.

He is currently Managing Director of the fish feed business, BioMar Ltd., and Research Coordinator for the Biomar Group.

Handbook of Trout and Salmon Diseases

Third Edition

Ronald J. Roberts

BVMS, PhD, FRCPath, FRCVS, FIBiol, FRSE
Emeritus Professor of Aquatic Pathobiology
Institute of Aquaculture, University of Stirling
Technical Director, Landcatch Ltd

and

C. Jonathan Shepherd

PhD, MSc, BVSc, MRCVS
Managing Director BioMar Ltd

Fishing News Books

Copyright © RJ Roberts and
   CJ Shepherd 1974, 1986, 1997

Fishing News Books
A division of Blackwell Science Ltd
Editorial Offices:
Osney Mead, Oxford OX2 0EL
25 John Street, London WC1N 2BL
23 Ainslie Place, Edinburgh EH3 6AJ
238 Main Street, Cambridge
   Massachusetts 02142, USA
54 University Street, Carlton
   Victoria 3053, Australia

Other Editorial Offices:

Blackwell Wissenschafts-Verlag GmbH
Kurfürstendamm 57
10707 Berlin, Germany

Blackwell Science KK
MG Kodenmacho Building
7-10 Kodenmacho Nihombashi
Chuo-ku, Tokyo 104, Japan

First published 1974
Second Edition 1986
Reprinted 1990
Third Edition 1997

Set in 10.5 on 12pt Palatino
by Excel Typesetters Co., Hong Kong
Printed and bound in Great Britain by
Hartnolls Ltd, Bodmin, Cornwall

DISTRIBUTORS

Marston Book Services Ltd
PO Box 269
Abingdon
Oxon OX14 4YN
(*Orders*: Tel: 01235 465500
      Fax: 01235 465555)

USA
   Blackwell Science, Inc.
   Commerce Place
   350 Main Street
   Malden, MA 02148 5018
   (*Orders*: Tel: 800 759 6102
               617 388 8250
         Fax: 617 388 8255)

Canada
   Copp Clark Professional
   200 Adelaide Street West, 3rd Floor
   Toronto, Ontario M5H 1W7
   (*Orders*: Tel: 416 597-1616
               800 815 9417
         Fax: 416 597 1617)

Australia
   Blackwell Science Pty Ltd
   54 University Street
   Carlton, Victoria 3053
   (*Orders*: Tel: 03 9347 0300
         Fax: 03 9347 5001)

A catalogue record for this title
is available from the British Library

ISBN 0-85238-244-8

Library of Congress
Cataloging-in-Publication Data
Roberts, Ronald J.
   Handbook of trout and salmon diseases /
   Ronald J. Roberts and C. Jonathan Shepherd.
   – 3rd ed.
      p. cm.
   Includes bibliographical references (p.)
      and index.
   ISBN 0-85238-244-8 (hb)
   1. Trout – Diseases. 2. Salmon –
Diseases. I. Shepherd, C. Jonathan.
II. Title.
SH179.T8R635   1997
639.3'757 – dc21   97-14109
                                    CIP

Contents

Preface

The first edition of this book was produced as a simple guide for trout and salmon farmers in the early 1970s. Since that time salmonid farming has grown beyond recognition; in particular, farming of salmon has become a major new industry in Norway, Chile, Scotland and elsewhere. At the same time our volume of knowledge on salmonid diseases has greatly expanded, as has also our practical understanding of husbandry and health management.

The first and second editions of the handbook have been well received by the fish farming industry over the past 20 years. We hope that this new, greatly expanded edition will be equally well accepted and of practical value on a day-to-day basis.

R. J. Roberts
C. J. Shepherd

Acknowledgements

This is the third edition of a handbook which over 20 years has evolved from a short practical text for use as an adjunct to short courses to a volume with over a hundred plates and a wide range of descriptions of agents of disease not even imagined in the salmonid farming industry of the 1970s. We have again to thank Dr Ted Needham in Canada and Chris Poupard in England for their original help in creating the first edition. We are also indebted to the large number of veterinarians and fish biologists who have kindly allowed us to use their often outstanding illustrations to embellish the text. We hope they are all acknowledged in the text and apologize for any omissions. Aquaculture has now become an established part of the world's food production capacity. We have been proud to be able to make a contribution to these developments and hope that the third edition of this handbook will prove useful to the new generation of fish farmers.

We originally worked together, at the University of Stirling, through the generous support of the Nuffield Foundation. A quarter of a century later we are very pleased to acknowledge their help. Their investment is still yielding dividends throughout the world.

Liam Kelly, John Springate, David Whyke and Rod Wooten all made valuable comments on the text of this latest edition. We also thank Christine Kerr, Ilse Marshall and Senga McLean for their secretarial assistance and, of course, our wives Helen and Maria, without whose forbearance this text would certainly not have seen the light of day.

List of Abbreviations

BKD	bacterial kidney disease
BOD	biological oxygen demand
EIBS	erythrocytic inclusion body syndrome
ERM	enteric redmouth
HCG	human chorionic gonadotrophin
IHN	infectious haematopoietic necrosis
IPN	infectious pancreatic necrosis
ISA	infectious salmon anaemia
OMV	*Oncorhynchus masou* virus
OSD	Oregon sockeye disease
PD	pancreas disease
PKD	proliferative kidney disease
PMS	post-viral myopathy syndrome
SRCD	Sacramento river chinook disease
TB	tuberculosis
TSA	tryptone soya agar
UDN	ulcerative dermal necrosis
VEN	viral erythrocytic necrosis
VHS	viral haemorrhagic septicaemia

Chapter 1
Cultured Salmonids

The farming, or culture, of fish under controlled conditions has been practised for thousands of years, but the farming of salmonid fish – the salmon and trout groups – is a fairly recent activity. Salmon and trout were first hatched and reared under artificial conditions during the last century, primarily for stocking waters for anglers, although there are some claims that the Red Indians of Western America practised a form of ranching culture by moving fertilized eggs from one river to a less well stocked one.

The Danes pioneered the farming of rainbow trout (*Oncorhynchus mykiss*) for human consumption, and now there is a significant rainbow trout farming industry in most of the cooler areas of the developed world. Farming of Atlantic salmon (*Salmo salar*) has grown very rapidly, particularly in Scotland, in Norway, and more recently in Chile and Tasmania. Pacific salmon (*Oncorhynchus* spp.) are native to the western seaboard of the USA and Canada and are cultured there and also in Chile and New Zealand. The culture of salmon takes the form both of closed culture, where all of the stages of the life cycle are carried out in captivity, and ranching, where the young fish are released in order to range the ocean for their feed, and return when mature for capture at the site of release.

Other salmonids regularly farmed include the brown trout (*Salmo trutta*) and the brook trout (*Salvelinus fontinalis*), both of which are farmed predominantly for release in angling waters for restocking purposes, and the Arctic char (*Salvelinus alpinus*).

The family Salmonidae is usually divided into two sub-families, the Salmonini, which includes all of the sport and commonly cultured species, and the Coregonini, which includes the whitefish or freshwater herrings (gwynniad, pollan, powan, vendace) (see Table 1.1).

This book is concerned solely with the Salmonini and when the term salmonid is used, it is as a collective noun embracing all four genera, a genus being a collection of related species.

1.1   GENUS *SALMO*

The most famous of the salmonids is the Atlantic salmon. This is closely related to the brown trout and sea trout. The young salmon is referred to as a parr until it develops its silvery coat, prior to migrat-

Table 1.1 The Salmonidae

Coregonus	*Coregonus nasus* (broad whitefish)
	Coregonus clupeaformis (lake whitefish)
	Coregonus lavaretus (powan)
Salmo	*Salmo trutta* (brown/sea trout)
	Salmo salar (Atlantic salmon)
Oncorhynchus	*Oncorhynchus masou* (Japanese salmon)
	Oncorhynchus kisutch (coho)
	Oncorhynchus tschawytscha (chinook)
	Oncorhynchus keta (chum)
	Oncorhynchus nerka (sockeye)
	Oncorhynchus gorbuscha (pink)
	Oncorhynchus mykiss (rainbow trout)
	Oncorhynchus clarkii (cut-throat trout)
Thymallus	*Thymallus thymallus* (grayling)
	Thymallus arcticus (Arctic grayling)
Salvelinus	*Salvelinus fontinalis* (brook trout)
	Salvelinus malma (Dolly Varden)
	Salvelinus namaycush (lake trout)
	Salvelinus alpinus (Arctic char)

ing as a smolt to sea, where it grows very rapidly on the rich food available there. Returning from sea, it is referred to as a grilse if it returns after one winter at sea, and a salmon if it has spent more than one winter at sea. It is often exceedingly difficult to distinguish between juvenile Atlantic salmon and brown trout, and between large sea trout and adult Atlantic salmon. Table 1.2 indicates the various differences which have been described, but no one feature is sufficiently constant for complete reliability.

The rainbow trout, which has recently been reclassified as a member of the genus *Oncorhynchus*, is usually distinguished by its smaller spots, its smaller scales, the more abundant spotting of tail and fins, and the iridescent line down each side which is evident in certain conditions of incident light.

In freshwater, rainbow trout grow much faster than Atlantic salmon or brown trout, and so for culinary purposes is the species most frequently cultured. Brown trout are almost exclusively farmed for restocking purposes. Atlantic salmon were originally reared solely up to the parr stage, to place in streams in order to increase future returning adult numbers, either for the net or for angling or else to replace fish which would have hatched in nursery streams impounded for hydro-electric reservoir purposes. However, they are now farmed in very considerable quantities in combined freshwater and marine facilities, for table production of adult fish.

Table 1.2 Differentiation between Atlantic salmon (*Salmo salar*) and brown trout (*S. trutta*)

Feature	*S. salar*	*S. trutta*
Juvenile fish		
Parr marks	10–12	9–10
Dorsal fin rays	10–12	8–10
Adipose fin	Black	Red tip
Ventral fins	Large and white	Smaller, orange-red
Tail	Deep fork	Shallow fork
Upper lip	Short, only to mid-eye	Long, to behind eye
Scales between lateral line and adipose fin	13–16	11
Adult fish		
Scales between lateral line and adipose fin	13–16	11
Wrist (tail peduncle)	Very slim (grilse) Slim (spring fish)	Stout and short
Tail	Swallow tail (grilse) Forked (spring fish)	Flat-ended
Scale reading	Wide winter rings; usually no spawning marks	Narrow winter rings; usually many spawning marks on a large specimen

The sea trout is considered to be identical to the brown trout – merely a brown trout which has decided to feed in the richer waters of the sea – and the steelhead is a similar form of rainbow trout. Like salmon, sea trout parr also smoltify and become silvery-coloured on going to sea.

There are a number of areas where land-locked Atlantic salmon occur, and these use large lakes as their 'inland sea'. The Atlantic salmon of the Baltic are almost exclusively produced in hatcheries, mainly in Sweden, to compensate for the loss of former nursery streams to hydro-electricity production or factory use, and this form of sea-ranching was one of the first to demonstrate what a useful form of production this can be, where homing takes place and provided international agreement allows. Attempts to ranch fish in the southern hemisphere have been unsuccessful.

1.2   GENUS *ONCORHYNCHUS*

Pacific salmon are found in the rivers of most north Pacific coasts, both east and west. In addition, they are sometimes found in Atlantic salmon fisheries, as a result of transplantations made from former USSR Pacific waters to north-west former USSR Arctic/Atlantic Ocean waters. Recently the rainbow trout, previously known as *Salmo gairdneri*, was reassigned to the genus *Oncorynchus* as *O. mykiss*.

The generic name of the Pacific salmon, *Oncorhynchus*, means hooked nose. This refers to the very marked hooked snout and kype possessed by most of the males at spawning. The duration of the different stages of the life cycle of Pacific salmon is more constant than that of the Atlantic salmon, and all Pacific salmon die after spawning.

The largest of all the salmon, the king salmon, quinnat or chinook salmon (*O. tschawytscha*), has been recorded weighing over 40 kg although the average is about 9 kg. Chinook have black mouths and gums and become progressively darker until they die after spawning. Chinook are cultured for replacement in rivers affected by hydro-electric schemes on the American and Canadian coasts. They are not extensively farmed for the table, although there is some production in Canada and Chile. The chinook often swims as much as 1500–2000 miles up the Yukon or Fraser rivers to spawn, and this migratory feat, while producing eggs and milt and without feeding, is a major functional reason for its large size on entering the river.

The coho salmon (*O. kisutch*) is probably the best Pacific salmon for the table, with its very red flesh and good smoking and freezing qualities. It is farmed in the former USSR, USA, Canada, Chile and Japan. A silvery fish in saltwater, the coho turns red when moving into freshwater to spawn. As well as the very prominent kype in the male, both sexes develop white nostrils.

The sockeye salmon (*O. nerka*), the most popular of the Pacific salmon for canning purposes, is known as the blue-back when at sea because of its bright greenish-blue hue, but on entering freshwater it turns very bright red, with a greenish head.

The pink (*O. gorbuscha*) and chum (*O. keta*) salmon differ from the other Pacific salmon in that their young migrate to sea immediately after hatching instead of having a freshwater feeding period. The chum is a largish (5 kg), slender fish which is very silver at sea, but develops red irregular blotches when coming into freshwater. The pink is much smaller, and is the species also found occasionally in the Atlantic as a result of the former USSR releases.

All Pacific salmon species have been investigated for their food farming potential, but only the farming of coho has developed into an industry of any significance, mainly in the USA and Chile.

Oncorhynchus masou is a species mainly found in northern Japan. It is very similar to *O. nerka*, the sockeye salmon.

There are land-locked varieties of both Atlantic and Pacific salmon, the Atlantic land-locked races being mainly found in eastern America and northern Europe, and the Pacific ones, notably the kokanee variety of the sockeye, in the western USA and Canada.

1.3   GENUS *THYMALLUS*

The grayling (*T. thymallus*) which is found in freshwaters of northern Europe spawns in summer like the coarse fish. It is the nearest member of the Salmonini to the other branch of the Salmonidae, the Coregonini, which includes certain whitefish. Although grayling have been cultured, they are usually considered deleterious to a trout or salmon fishery.

1.4   GENUS *SALVELINUS*

The members of the genus *Salvelinus* are known as the chars; (in the USA many of the species referred to as trout are in fact chars).

The characteristic distinguishing feature of the chars is the vomer bone which is the longitudinal bone in the top of the mouth. In the trout and salmon it is flat and covered with teeth all the way down, whereas in the chars it is keel-shaped, and has very few teeth and only at the top. The chars are usually redder in colour than salmon and their spots are usually lighter in colour than the background.

Chars are not usually cultured for the table, but the American varieties such as the brook trout, the Dolly Varden trout and lake trout are cultured in the USA and Canada and in some European countries for restocking. Normally an inhabitant of colder waters, the Arctic char is only found in deep lakes, usually at high altitudes.

The chars feed at lower temperatures than rainbow trout or brown trout. In very cold climates, therefore, they may be better capable of weight gain than even the rainbow trout, which grows very rapidly at higher temperatures.

Chapter 2
Anatomy of Salmonid Fish

Salmonid fish are well adapted to their lifestyle, being fast and sharp-eyed to catch their prey. Since they eat live food they have the typical short digestive tract of the carnivore. Many of the diseases of salmonid fish, or their signs, are directly related to the anatomy of the fish. The following descriptive chapter is therefore essential to allow the reader to understand the significance of the sites of disease in salmonids.

The anatomy of a typical salmonid is depicted in Fig. 2.1 and described under the organ systems, as follows:

(1) The integumentary system – skin and appendages.
(2) The muscular system – muscles and bones.
(3) The respiratory system – gills.
(4) The circulatory system – heart and blood vessels.
(5) The digestive system – mouth, stomach, intestine and associated glands.
(6) The excretory system – kidney and bladder.
(7) The reproductive system – ovaries and testes.
(8) The nervous system – brain, spinal cord and nerves.
(9) The endocrine system – pituitary, adrenal and other hormone-producing glands.

2.1   THE INTEGUMENTARY SYSTEM

The skin of the salmonid is essential for waterproofing the fish as well as providing the armour plating of the scales. This function of keeping the water out, and fish tissue-fluids in, is performed by the epidermis. This is a very delicate clear covering which is draped over the scales and possesses small glands, the goblet cells. These increase in number when the fish is stressed, and help to secrete the slimy mucus, a protective coating of infection-resistant slippery fluid, which also makes the fish difficult to hold. Sometimes skin parasite infections cause the epidermis to secrete a thicker, more viscid mucus, which gives it a bluish tinge, but usually it is clear.

Beneath the epidermis lie the scales – ovoid plates of bony material which are formed in small pockets or scale beds. Scales develop on salmonids at the fry stage, and once a fish has its full quota of scales, it does not develop more as it increases in size. Consequently, the

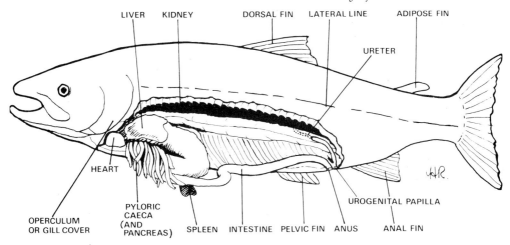

Fig. 2.1 Anatomy of a typical salmonid fish.

scales must grow in step with the fish. Scales grow by accumulation of material round their edges, laid down in the form of concentric rings. When a fish is growing rapidly these rings are spaced far apart so that in the summer (or in the sea) the distance between rings is much greater than in winter. At spawning time, the salmonids do not feed, and in order to obtain sufficient calcium for eggs or sperm they withdraw calcium from the outermost scale rings. This results in permanent scarring of the scale at that place (Fig. 2.2). By examination of scales, salmon biologists are able to assess the age, number of spawnings, and even the fish's length at the end of each year of life. Occasionally a scale may be damaged and a new scale grows in the scale pocket. This scale cannot recapitulate the previous life history of the fish and that area of the new scale is therefore blank.

Over the scales are the pigment cells – melanophores or black cells, iridophores or silver cells, and xanthophores, which produce the yellow and red spots. The black cells are under both nervous and hormonal (i.e. chemical) control. When fish are on a dark background they emphasize their black pigment cells, and on a light background the silver cells are more obvious. When they are depressed due to disease, fish frequently become darker in colour.

The strength of the skin is in the dermis – the layer below the scales. This is a very fibrous layer with considerable tensile strength.

2.2   THE MUSCULAR SYSTEM

The majority of fish muscle, the muscle that we eat, is known as white muscle. This is a very powerful type of muscle, used principally for escape, or pursuit of prey. Red muscle, which is similar to the muscle of higher animals, is only found along the lateral line and in some specialized sites such as the base of the fins, the eye and gills.

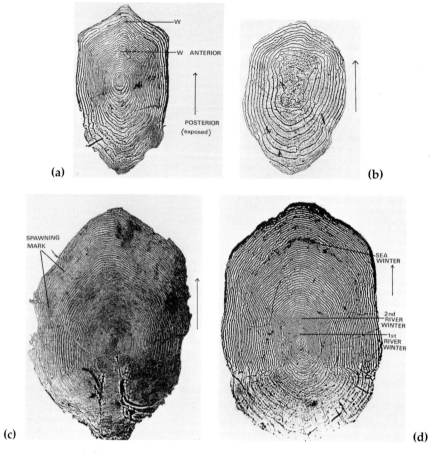

Fig. 2.2 Scales from Atlantic salmon at different stages, showing features of the life cycle which may be determined by scale reading. (a) Scale from a two-year-old smolt showing two winter growth periods (W) separated by a summer band, with a second summer band with wide rings being laid down on the margin. The posterior part of the scale is rougher and more worn than the anterior, which is protected within the scale pocket. (b) A replacement scale formed after loss of the original scale. The central empty zone is scar tissue surrounded by normal circuli. (c) Adult salmon scale showing a spawning mark, where calcium has been resorbed from the scale. (d) Grilse scale showing two years of river life (scale (a)) and one year in the sea. This is the normal pattern for most grilse in rivers. (*Courtesy of Dr L.M. Laird.*)

The main swimming muscles of salmonids are arranged in a series of blocks of white muscle or myotomes. This gives them considerable driving force on the tail. The myotomes are attached to the spine, the central bone, which is very flexible. The fins are moved by small independent red muscles. In addition to the usual fins, the salmonid fish are all characterized by a small appendage on the back, just in front of the tail, which is known as the adipose fin because of its fatty internal structure.

The skeleton of the young fish, as with other animals, is formed of cartilage, which becomes calcified later and can have an important influence on certain diseases.

2.3   THE RESPIRATORY SYSTEM

Fish breathe by means of gills, a system of four sets of very fine flattened capillaries or tubes, on either side of the throat, through which the blood flows and over which water is continually passed. In passing through the gills, the blood gives up its carbon dioxide to the water, and obtains oxygen from the water, through the gill wall. The respiratory surfaces of the gills, the secondary lamellae, have to be very delicate so that the oxygen and carbon dioxide can be readily exchanged. They also contain mucus-producing cells and cells that excrete any excess salt from the blood as it passes through them, and they also excrete ammonia. Obviously such a delicate structure on the outside of the body is highly vulnerable to injury via the water. It is also likely to be seriously compromised if it becomes a site for bacterial or parasitic infection. The gills are protected on the outside by a bony shield called the operculum, and on the inside of the throat they have a set of comb-like structures called gill rakers, which help to guide the food down to the gullet rather than over the gills (Plate 1).

2.4   THE CIRCULATORY SYSTEM

The circulatory system is the blood-transport system of the fish. The pump in the system is the heart, a muscular organ occupying the area at the base of the throat. It is a two-chambered pump (lacking the auxiliary pump for taking blood to the lungs, which is a feature of man and the higher animals). The blood passes from the triangular, very muscular ventricle, which provides the main pressure, into the white, elastic-walled bulbus arteriosus (Plate 2). This is an elastic pressure-balance converting the pumping of the heart into a steady surge of blood to the gills, from whence it passes to the rest of the body to deliver up oxygen to the tissues. Once it has passed through the gills its pressure is much reduced and its passage through the tissues is relatively slow. In the fine circulatory network of the tissues, the capillaries, the oxygen of the blood is replaced by carbon dioxide and waste products. The blood then returns via the vena cava or great vein, passing through the kidney on its way back to the heart. As the blood passes through the capillaries, some fluid, known as lymph, is lost to the tissues. This is the watery fluid which runs from a fresh fillet of fish. The lymph is returned to the circulation by a separate set of vessels, the lymphatics, which return it to the blood-stream just before the heart.

2.5 THE DIGESTIVE SYSTEM

The digestive system consists of a relatively simple tube in the salmonids (Fig. 2.3). It starts at the mouth, where the teeth are designed for capture not chewing. When ingested the food is quickly passed down the gullet or oesophagus to the stomach, a U-shaped organ which can expand greatly to take large meals. It is in the stomach that the food is really chewed, i.e. it is broken down by the action of acid and digestive enzymes as well as the crushing contractions of the muscles in the wall of the stomach. At the posterior end of the stomach where it joins the small intestine, there is a group of blind-ending sacs, the pyloric caeca. These usually number 30–80, and they lie conspicuously across the stomach when the fish is opened. They are covered with a considerable amount of white fatty tissue unless the fish has been starved. From the stomach, food passes through a one-way valve, the pylorus, to the intestine, where the disintegrated food is acted upon by further enzymes. These break down the food to its constituent sugars, fats and amino acids (from proteins), which then pass into the bloodstream of the intestinal wall for transport to the liver. The remaining food – roughage, snail shells, etc. – travels on to the large intestine and is voided as faeces.

Associated with the digestive tract are two very important glands. One, the liver, is a large organ situated just in front of the stomach. It

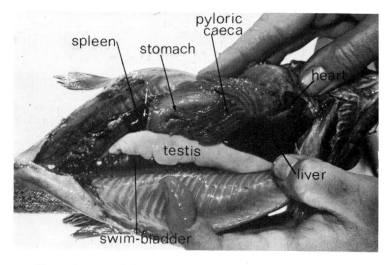

Fig. 2.3 Ventral view of the abdomen of a rainbow trout at post-mortem. The abdominal cavity should be full, but without excessive fluid in it. The liver should not be pale or bronzed and there should be a small amount of fat around the pyloric caeca.

is a pinky-brown colour, soft and easily ruptured. The liver is the main factory of the body, to which food molecules are taken in the blood from the intestine for manufacture into the proteins, carbohydrates and fats of the fish's body. Inserted in the top of the liver is a small greenish sac – the gall bladder. When incised, this usually releases a greenish fluid called bile, which under normal conditions passes to the intestine through the bile duct and aids with food breakdown. Because of its importance in food metabolism, disease of the liver is very significant. The commonest liver abnormalities are excessive infiltration by unsuitable dietary fats and parasitism. Parasites are also frequently found in the gall bladder.

The other important associated digestive gland is the pancreas. This is a very diffuse structure which cannot be seen with the naked eye as it is scattered throughout the fat surrounding the pyloric caeca. The pancreas has two functions: the production of pancreatic enzymes, which pass via the pancreatic duct to the intestine, and the production of insulin, which controls sugar and protein metabolism and prevents fish from becoming diabetic. The pancreas is very significant in virus diseases because it is a favourite site of multiplication for two of the most important salmonid viruses.

Many species of fish possess a swim bladder, which is a hydrostatic organ used to trim buoyancy at the appropriate depth (Fig. 2.4). The swim bladder may also have a function as a hollow organ for receiving deep, low frequency sounds. In the salmonids it has a connection with the back of the throat so that the fish can quickly

Fig. 2.4 Swim bladder. This illustration shows the abdominal cavity with the majority of the viscera removed. The swim bladder may be dilated but should not contain excess fluid or be thickened and it should not be tightly filled with gas. The kidney is immediately below the swim bladder in this picture.

squeeze out air and drop to the bottom. 'Any blockage of this duct, which can take place with dusty food, or damage to the swim bladder wall, can result in considerable swimming problems for the fish.

2.6   THE EXCRETORY SYSTEM

The kidney is the main filter of the body. It filters blood through a sieve-like apparatus called the glomerulus and passes it through tubes to paired ducts, the ureters, which carry it to the bladder. In salmonids the bladder is a small thin-walled structure above the anus. The duct from the bladder drains via the urogenital opening, which is also the exit for eggs (Plate 3).

The kidney of the salmonids is a long black structure in the top of the abdomen, extending from the back of the head to the vent. The vena cava runs through the centre of the kidney and on its outer surface may be seen the narrow white ureters, wending towards the bladder.

In higher animals the kidney is purely a selective filter, but in fish it also contains the haemopoietic tissue, especially at the front end of the kidney. This is the tissue which makes the oxygen-carrying red blood cells and defensive white blood cells, and also stores them until they are needed. The other site where this takes place is the spleen, a large black organ attached to the wall of the intestine. This haemopoietic tissue of kidney and spleen is very important in disease as it is affected by a number of serious bacterial and viral agents. It also contains a network of traps, the fixed macrophage cells, which catch any microbes passing through in the bloodstream and usually succeed in destroying them.

2.7   THE REPRODUCTIVE SYSTEM

The gonads of the salmonids comprise paired ovaries in the female and testes in the male. In the immature or resting state they lie in the anterior of the abdomen, above and on either side of the stomach. At sexual maturity, under the control of the hormones from the pituitary gland, they develop to extend the full length of the abdomen. The ovary consists of germinal cells, some of which grow to the size of a pea to form the orange-coloured ova or eggs. Others stay small as the cells for subsequent spawnings.

At spawning the skin of both male and female becomes thicker and more shiny, and also the urogenital opening swells up. Eggs are released into the abdomen as the supporting capsule ruptures, and are pushed on a tide of fluid to the urogenital opening by contractions of the female's abdominal muscles and by small sweeping hair-like cilia inserted in certain parts of the lower abdominal wall.

Semen, known as milt in fish, is excreted from the testes by bodily contraction and passes into the water as a cloud of living, wriggling sperm cells. In the wild this occurs in the redd prepared by the female, but hatcherymen assist the process artificially.

2.8   THE NERVOUS SYSTEM

The nervous system of salmonid fish reflects their behaviour. Salmon home on their sense of smell, hunt with their eyes, and are creatures almost entirely of reflex. Consequently they have well developed olfactory areas at the front of the brain, which connect directly with the nostrils (Plate 4). These are paired sacs on the snout and have a continual flow of water over their sensitive tissue. The area of the brain responsible for sight (the optic lobes) comprises two large rounded structures occupying almost a third of the brain's volume, whereas the cerebral tissue (corresponding to the massive cerebral hemispheres, or thinking area, of the human brain) comprises two very small areas on the side of the brain stem (Fig. 2.5). The brain is extended backwards, as the spinal cord, from which arise nerves to muscles, organs and skin. The spinal cord passes down the middle of the vertebrae of the spinal column, which gives it considerable protection. The skin contains certain nerve endings which can detect touch and pain. The most interesting of the surface sensory structures are the lateral lines – symmetrical, long, narrow, fluid-filled tubes

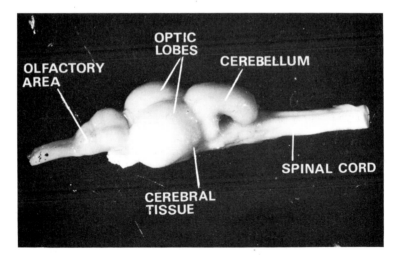

Fig. 2.5 Brain of an Atlantic salmon. The salmonids are hunters with keen eyesight and a good sense of smell. Hence the front part of the brain, which is concerned with smell and is continuous with the nostril, is quite well developed, and the paired optic lobes of the brain (the large domes) comprise its largest component. The thinking part, or cerebral cortex, is very small and confined to just below and behind the optic lobes. The cerebellum is responsible for the continuous coordination of information received via the spinal cord from various parts of the body regarding position and activity.

extending from the back of the head to the tail, with pores opening to the exterior, and sensitive nerve endings lining their inside wall. The lateral line is believed to receive low frequency pressure waves in the water.

The eye of the salmon, as with all fish, has no eyelids, but internally is similar to that of other animals. The lens is round and completely clear in the normal fish, but in certain disease conditions becomes cloudy. The eye is also one of the sites with extremely delicate blood vessels and is therefore very vulnerable to rupture of capillaries, e.g. by gas bubbles in certain circumstances.

The ears of salmonid fish do not have an outlet and although they may detect some vibrations, their main function is balance. They are located within the skull, just behind the eyes, and when they are damaged by disease the fish is unable to balance properly.

2.9   THE ENDOCRINE SYSTEM

The endocrine glands are small groups of cells which have a significance for the body way beyond their size. They secrete chemicals into the bloodstream, hormones, which act on distant sites such as the gonads, skin or blood vessels. The most important endocrine gland, the pituitary, which has been called the 'conductor of the endocrine orchestra,' is located in a very secure site below the brain.

The adrenal, or inter-renal as it is often called in fish, is a gland producing several important hormones, including the fear hormone adrenaline. It is located within the haemopoietic tissue at the anterior end of the kidney. The thyroid gland, producing growth hormone, is elusive in the salmonids, often scattered randomly around the tissues of the throat area. Salmonids and other fish also have two endocrine structures with as yet unknown functions. These are the Corpuscles of Stannius, small white spots, placed laterally in the mid-kidney tissue (usually three or four can be seen on the surface of the kidney), and the pseudobranch, a red, vestigial gill-like structure on the inside face of each operculum. The pseudobranch has been associated with an endocrine hormone function, but recent evidence suggests it is more likely to be associated with ensuring that high levels of oxygen reach the brain at all times and also more general maintenance of oxygen and carbon dioxide levels in the blood.

There are three other endocrine glands of fishes, which do not have equivalents in higher animals. These are the ultimobranchial glands concerned with calcium metabolism, the urophysis, which is a swelling near the end of the spinal cord, and the pineal, at the top of the head, which is thought to be light sensitive and to be associated with pigment cell control and the seasonal spawning cycle. The ovary and testis also have an endocrine function, producing sex hormones.

Chapter 3
Husbandry of Salmonid Fish

Successful farming of salmon and trout relies upon a high standard of husbandry and stockmanship, which depends in turn on understanding the fishes' biological requirements for survival and growth. Fish farming has many features in common with conventional livestock production and the good stockman needs to provide suitable food and housing. Being cold-blooded, fish simply adopt the temperature of their aquatic environment. Just above freezing, salmon and trout are lethargic and expend very little energy apart from what is needed for their basic bodily functions or metabolism. As the water temperature increases so does their activity level and hence need for food and oxygen in order to provide the necessary energy. Within limits, faster growth of fish occurs at higher temperatures, but unfortunately this is true also for the microorganisms that cause fish diseases. However, at temperatures much above 25°C, the water does not contain enough oxygen to satisfy the normal metabolic requirements of salmon and trout, which therefore suffocate. So fish husbandry is closely linked to the chemical and physical factors involved in water quality.

Fish are conscious animals and responsible farm management and husbandry systems should take this into account. The principles of land animal welfare therefore also apply to farming of trout and salmon. This means that:

- farmed fish should not be subjected to hunger or malnutrition;
- they should be free from discomfort (e.g. reared at the correct temperature and stocking densities);
- they should be free from pain, injury or disease (e.g. correct handling methods and equipment designed to minimize stress);
- they should be free from fear and distress (e.g. minimize risk of attacks by predators);
- they should be free to express normal patterns of behaviour;
- they should be free from stress or suffering when transported (e.g. reduction of water temperature and oxygenation of water in transport tanks);
- they should be free from stress or suffering when slaughtered (e.g. humane instantaneous procedure instead of allowing fish to suffocate out of the water).

Most salmonid farming industries now have codes of good farm practice and it is important that fish farmers adopt their recommendations to ensure that good welfare is achieved in day-to-day practice.

3.1   ENVIRONMENTAL REQUIREMENTS

3.1.1   Oxygen

The minimum level of dissolved oxygen required by trout and salmon is about 5.5 mg/l and about 7 mg/l for salmonid eggs. The ability of water to take oxygen into solution is, however, governed by temperature, pressure and its dissolved salts. The higher the temperature, the less oxygen is retained in water. For example, in freshwater at 5°C and 1 atmosphere (i.e. sea level) of pressure, oxygen solubility is 12.8 mg/l; at 20°C oxygen solubility is 9.2 mg/l. If atmospheric pressure is reduced, solubility is also reduced and this may cause problems for fish farms at very high altitudes. Also, as the salinity is increased, solubility is again reduced, the oxygen solubilities in freshwater and full strength seawater at 10°C and 1 atmosphere of pressure being 7.9 mg/l and 6.4 mg/l respectively. The interrelationships between water temperature, salinity and the dissolved oxygen content are shown in Fig. 3.1. However, it is important to note that these values assume the water is fully saturated with oxygen which is frequently not the case in practice, hence it is also important to measure the actual amount of oxygen present. To try to ensure that the incoming supply of water is fully saturated with oxygen, it is advisable to install splash-boards at the inlet of each

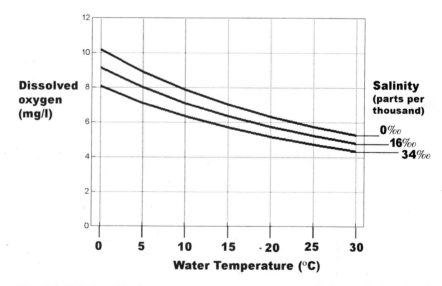

Fig. 3.1 Relationship between water temperature, salinity and dissolved oxygen. (Source: Parsons *et al.*, 1984.)

pond and to install aerators for use in danger periods (high temperatures, medication, etc.) or where stock levels are high, as in high
intensity smolt production.

In fish farms, as the temperature rises, the fish are usually fed more
heavily so they will grow faster, and their rate of oxygen consumption is thus markedly increased. At the same time faeces, and perhaps
unused feed, build up on the bottom, unless self-cleaning is very
efficient, and decompose. During the hours of darkness, any aquatic
plants present, e.g. algae, will also be using up oxygen. So, even if the
incoming water is fully saturated, under conditions of heavy stocking rates and high temperatures, the amount of dissolved oxygen
present, particularly in ponds, may drop to dangerously low levels.
Rates of flow and depth of water must be adjusted so that the consequent water exchange rate is adequate for the density of fish stocked,
water temperature and feeding rate. Care should be taken to ensure
that calculations of the usable volume of the pond exclude areas of
stagnant water and to remember that fish consume oxygen at a
higher rate when they are small, i.e. 1 kg of fry needs more water than
1 kg of growers. These factors highlight the need to monitor dissolved oxygen levels regularly with an oxygen meter, particularly at
high temperatures and heavy stocking rates.

3.1.2 Temperature

In terms of energy efficiency, fish farming is superior to the farming
of warm-blooded stock since fish, being cold-blooded, do not need to
expend energy in maintaining their body temperature independently
of the environment. Salmon and trout are coldwater fish and thrive
within the temperature range 0–25°C. Above about 20°C, oxygen
levels become limiting as the dissolved oxygen concentration of water falls with rising temperature and it is necessary to starve the fish
in order to minimize their oxygen requirements. Incubation of
salmonid eggs is best carried out at temperatures below 13°C,
whereas optimum fish growth occurs at around 16°C depending on
the species.

In order to obtain optimum and uniform growth, many hatcheries
now heat the incoming water to a consistent 10–15°C. This allows
good growth, but care must be taken when transferring such fish to
ambient systems for on-growing since they are vulnerable to short-
term temperature fluctuations until acclimatized.

3.1.3 pH

The pH value of the water is a measure of its acidity or alkalinity and
is measured on a scale whose range is from 0 (very acid) to 14 (very
alkaline) with the neutral point at 7. Since it is a logarithmic scale, a
change of 0.1 pH units represents a marked change in acidity. It is
important that the pH should not fluctuate but remain stable at some

point within the approximate range 6.4–8.4, and pH may easily be measured by portable meters. Variations in pH stress the fish and are less likely to occur if the water is buffered, as in limestone areas. This is because such water contains large amounts of dissolved salts, especially calcium salts, which resist changes in pH and which favour natural plant and animal life. Acidic water is found in regions deficient in calcium and associated with igneous rock. Very acidic waters may arise due to mineral acids leached out of the soil by spates. The consequent low pH can cause bleeding (haemorrhages) on the gills and heavy mortalities. In addition, low pH associated with acid rain, especially in run-off from coniferous forests, can also result in release of toxic levels of heavy metals, such as manganese and aluminium, from soils into the water. It is possible in such circumstances to add lime (calcium carbonate) to the inflow water in order to increase resistance to changes in pH. Unfortunately, there is increasing evidence of industrial pollution causing so-called acid rain which results in irreversible acidification of the terrain onto which it falls (see Chapter 11).

3.1.4   Ammonia

Ammoniacal substances are produced as excretory products by salmon and trout and are usually in a mixture of two forms: free ammonia (NH_3) and ionized ammonia (NH_4^+). The free ammonia is by far the more toxic of the two substances and concentrations as low as $0.02\,mg/l\,NH_3$ may be harmful to fry, causing damage to the gills and reduced growth, especially if dissolved oxygen levels are low. The relative proportions of free ammonia and ionized ammonia are dependent primarily upon the pH of the water, and the higher the pH, the greater the proportion of free ammonia, i.e. the toxic form. Therefore, although alkaline water has the great advantage that it is usually stable in pH, it has the disadvantage that any build-up in wastes is likely to be more dangerous than in acidic water. The prudent fish farmer should therefore monitor ammonia levels, particularly under conditions of alkaline water, low oxygen levels and high temperature. Figure 3.2 shows how the free ammonia levels vary with different pH and temperature values.

3.1.5   Toxic effects

It is by now obvious that fish continually excrete substances that are potentially harmful to themselves. Accumulation of faeces encourages the build-up of ammonia and also increases the amount of suspended matter in the water. Fish also excrete carbon dioxide and there is some evidence that this can cause kidney problems if it is not flushed away at high stocking densities. Suspended solids can harm fish in two ways: by mechanical irritation and clogging of the gills, and by removing oxygen from the water during decomposition. This

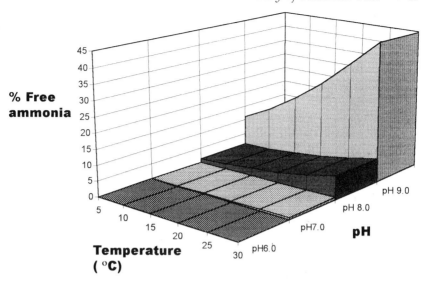

Temperature (°C)	% of total ammonia as free ammonia			
	pH 6	pH 7	pH 8	pH 9
5	0.01	0.13	1.24	11.18
10	0.02	0.19	1.83	15.70
15	0.03	0.27	2.68	21.59
20	0.04	0.40	3.83	28.47
25	0.06	0.57	5.44	36.53
30	0.08	0.81	7.52	44.84

Fig. 3.2 Variation of free ammonia levels with pH and temperature (after Kelly, 1997).

may be a problem when organic matter enters the water supply following a storm. In a fish farm it is very important to maintain an adequate water exchange rate and to clean the tanks and ponds regularly in order to prevent accumulation of waste material (this is generally not practical with earthen ponds without a lining where, however, a degree of self-cleaning may exist due to bacterial action of the mud bottom).

A particular water supply may therefore have insufficient oxygen for fish health due to its removal from solution by the demand of waste and organic material and accompanying microbes, etc. This is measured as biological oxygen demand (BOD); the oxygen consumed in milligrams per litre over a specified time by the wastes. The main reason that water pollution due to sewage or silage liquor frequently causes heavy fish mortalities is that the resulting high BOD level simply asphyxiates fish. However, water drawn from a spring, borehole or well may have insufficient oxygen due to lack of surface oxygenation from the air. Such water may also have a supersaturation of other gases (notably nitrogen) and this can cause disease in fish even if the oxygen levels are adequate. Occasionally ground water may also contain gases, such as hydrogen sulphide, which are toxic to fish. These problems can usually be adequately controlled by artificial aeration of the water supply before it enters the farm. By this means excess nitrogen and other gases may be blown off and the water saturated with oxygen.

Heavy metal ions may occur in water due to factors such as metal ores in the rock strata, the composition of pipes, or the presence of industrial effluent. These comprise ions such as iron, lead and copper which, except at very low levels, are toxic to fish. Iron salts can be a particular problem and may coat the surface of both eggs and gills causing asphyxiation and heavy losses. In addition, many substances are extremely toxic to fish and may enter the water supply accidentally, e.g. sheep dips and weedkillers, and even maliciously in the case of cyanide. The chlorine contained in much potable water is toxic to fish and most of the chemicals used for treating fish diseases have some toxic effects. Often they also remove some oxygen from the water, which imposes additional stress.

Finally, it should be borne in mind that ultraviolet radiation can cause harmful sunburn to fish, and hatchery tanks may need to be covered to prevent this becoming a problem under conditions of strong sunlight. Even ponds and raceways in the highlands of the tropics or in the southern hemisphere, where ultraviolet levels are high due to reduction in the ozone layer, may need some form of shade.

3.2 FOOD AND NUTRITION

Fish need energy in order to stay alive, swim about, grow, reproduce, etc. and this is supplied via the food. When digested by trout the three basic constituents of protein, fat and carbohydrate have been estimated to supply approximately 4, 8 and up to 4 kcal, respectively of available energy per g of food. Using protein to provide energy is much more expensive than using fat or carbohydrate and the fish farmer aims instead to convert as much as possible of the dietary protein into fish flesh. Salmon and trout are carnivorous fish in the wild environment, so it is hardly surprising that, under farm conditions, they require high protein diets, a fair proportion of which is usually supplied by fish meal. This is because the cheaper vegetable proteins do not contain all the amino acid building blocks essential for salmonid fish. In a similar fashion the dietary oils supplied for energy must be of a type containing the particular fatty acids that are essential to salmonids, usually fish oils. The analysis of typical compounded salmonid diets may be in the range as follows: protein 40–45%; fat 20–30%; carbohydrate 10–20%; water <10%. However, this says nothing about the digestibility of the various constituents, the available energy (i.e. calorific value), or the efficiency of conversion from food into flesh. It also ignores the key importance of providing essential vitamins and minerals without which growth will cease and deficiency diseases will ensue, perhaps with fatal results. Other additives include carotenoid pigments if the farmer is producing pink-fleshed salmon or trout.

In general the fish farmer aims for as fast a growth rate as possible at the minimum cost. Since use of low cost ingredients is likely to be ineffective and polluting, it is important to achieve efficient conversion of costly food into flesh with minimum waste for economical production. Figure 3.3 shows how both growth rate (live weight gain – LWG) and food conversion rate (FCR) of trout alter with age, the food conversion rate being expressed as follows:

$$FCR = \frac{\text{weight of food consumed}}{\text{increase in weight of fish}} = \frac{1}{\text{food conversion efficiency}}$$

The graph in Fig. 3.3 refers to experimental conditions, and FCR would be slightly higher under commercial farming conditions, but in both cases the conversion of food into fish becomes progressively less efficient with increased size. For example, trout reared to 200 g on high protein pellets might have an overall FCR of 1 kg of pellets required to produce 1 kg of trout. If their FCR was calculated as fry, it might be only 0.7 because pellets contain much less moisture than trout fry (the farmer is not really getting something for nothing).

Shortly after transfer to sea, young salmon have a FCR of approximately 1.0 using compounded feed, but thereafter it increases much in line with trout of the same weight until the onset of sexual maturity when FCRs of both trout and salmon can increase markedly. When calculating FCRs, the weights of any mortalities should also be recorded and stated, or included in the sum. It is only by careful attention to FCR that farmers can evaluate whether they are feeding correctly. Incorrect feeding rates are likely to cause poor conversion rates due to both physical wastage and consequent nutritional and disease problems. For example, too little food will cause losses from starvation, cannibalism and parasitism. Too much food will tend to foul the water with unused feed, removing oxygen and predisposing to various infectious diseases. Poor conversion efficiency is a common symptom of disease problems in fish, particularly where the disease is a slow insidious process of reduced growth, rather than a sudden 'acute' flare-up with heavy mortalities. Such 'chronic' disease states may pass unremarked if the poor conversion efficiency, i.e. high FCR, is not observed and investigated.

Feeding rates are commonly computer controlled and based upon water temperature and fish weight. Feed may be given to the stock by hand or by automatic feeders operated either by time switchgear or by the fish themselves. Hand feeding allows the farmer to observe

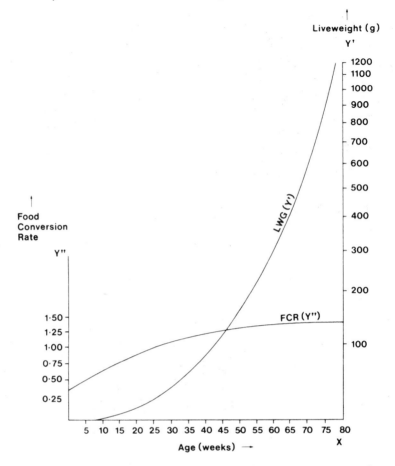

Fig. 3.3 Relationship between growth and food conversion rates and age of fish.

the behaviour of the stock. Self-feeders ('demand' feeders) allow the fish to regulate their own needs and are used on some trout farms. Trout seem unable to avoid overfeeding when the temperature is high, although it is probable that maximum food conversion efficiency is attained at a daily intake considerably less than that required to satisfy fish (i.e. *ad libitum* feeding). A problem with self-feeders is that more frequent grading may be necessary, as they seem to result in increased variation in size among a population; this may be due to a pecking order becoming established around the feeder. Giving the ration equally throughout the day rather than all at once is facilitated by automatic feeders, and may assist conversion efficiency, especially with fry, which should be fed up to eight times a day. It also helps to spread out daily oxygen requirements, and therefore is more economical on water.

Plate 1 Respiratory system of a rainbow trout. The operculum, or gill cover, is raised in this illustration to show the gills on their cartilaginous arches. The white 'teeth' on the inner gill arch are the gill rakers. Water is continually passing over the gills, bringing essential oxygen, but they are favoured sites for parasites as they provide shelter and a rich food supply (Courtesy of G. Macgregor.)

Plate 2 Heart of an Atlantic salmon. The heart is the main pump of the body and the large conical ventricle, with its coronary vessels is readily seen here. The white structure is the bulbus arteriosus and the dark area the atrium or auricle. (Courtesy of G. Macgregor.)

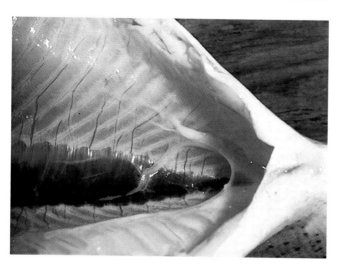

Plate 3 Kidney and urinary bladder. The black structure is the excretory kidney, within which the haemopoietic tissue is embedded. The bladder is the small white structure on the surface. (Courtesy of G. Macgregor.)

Plate 4 Nostril of Atlantic salmon. The covering flap has been removed to reveal the red sensory structures. (Courtesy of G. Macgregor.)

Plate 5a Two attached *Costia* cells on the developing gill of a salmon alevin.

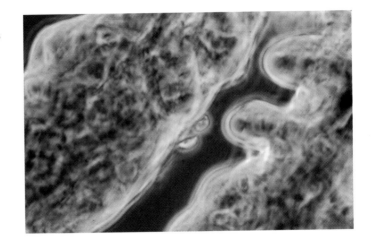

5b *Costia* attached to the gill of a trout fry. (Courtesy of C. H. Aldridge; copyright Unilever Research Laboratories.)

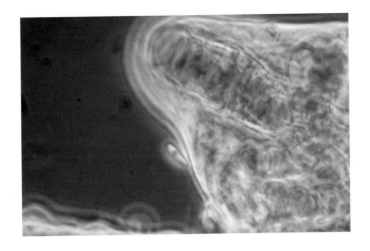

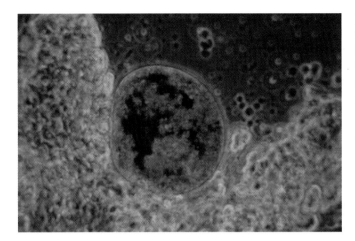

Plate 6 (a) Skin scraping from rainbow trout showing *Ichthyophthirius* cell having just invaded the skin, prior to developing its distinctive horseshoe-shaped nucleus.

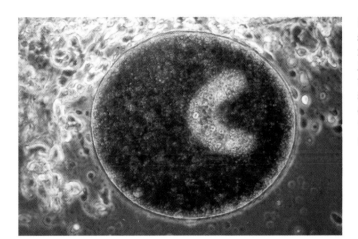

(b) Skin scraping showing adult mature *Ichthyophthirius* cell with horseshoe-shaped nucleus and dark colour typical of the adult. (Courtesy of G. Macgregor.)

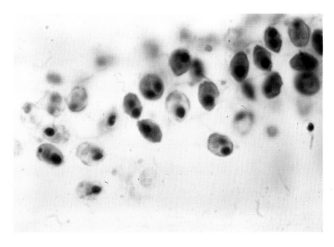

Plate 7 Whirling disease. This is a smear of chopped-up cranial cartilages stained with methylene blue of a fish dying of whirling disease. The *Myxobolus* (*Myxosoma*) spores characteristic of the disease show a shiny edge with two eye spots or polar capsules, and are best seen under the 40× objective.

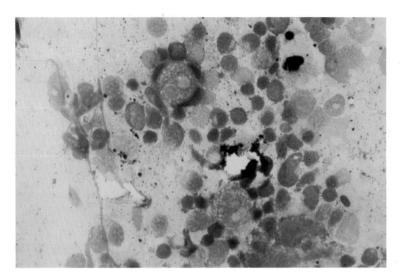

Plate 8 (a) Smear from kidney of PKD infected fish, showing kidney cells and large bluish multinucleate parasite cells.

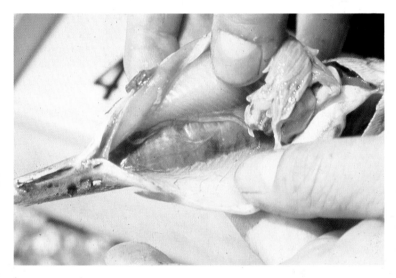

(b) Kidney of rainbow trout in acute stage of PKD. The kidney is grossly swollen and light in colour.

(Courtesy of Dr R. H. Richards.)

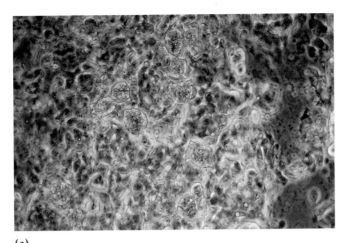

(a)

Plate 9 *Scyphidia* complex. These suctorial parasites are readily seen in skin smears of affected fish. The different members of the complex may have stalks, branches, or may be simple vase-shaped structures. Smears containing scyphidians usually also contain considerable cellular debris.
(a) Scyphidians on smear of salmon skin.
(b) Higher power magnification showing parasite and cellular material.
(c) High magnifica-tion of two parasites showing oral region and basal sucker.

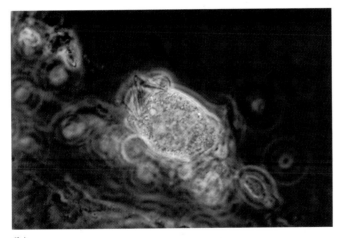

(b)

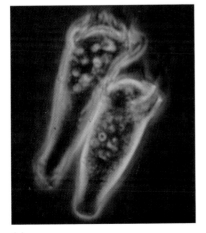

(c)

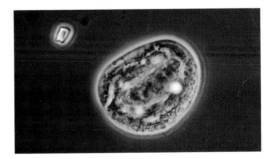

Plate 10 *Chilodonella*. This heart-shaped para-site moves slowly and is readily seen under low power magnification.

Plate 11 *Eubothrium* tapeworms removed from cut pyloric caeca and stomach of marine Atlantic salmon. (Courtesy of C.L. Oman.)

Plate 12 (above) *Triaenophorus* plerocercoid encysted in the liver of a rainbow trout. (Courtesy of Dr T. Håstein.)

Plate 13 (right) *Eustrongylides*. This red-coloured larval nematode is frequently found among the viscera of wild trout.

3.3 CYCLE OF FARMING OPERATIONS

Natural spawning in the northern hemisphere generally occurs from October to February, depending on the species. However, new broodstock management systems can modify this, and eggs taken from the southern hemisphere are also used to produce out-of-season fry. Under farming conditions, the male and female fish are manually stripped of their milt and ova which are mixed together. After fertilization has occurred, the 'green' eggs are placed in incubation systems. Development and hatching time are related to water temperature, and daily attention is necessary to ensure that dead eggs do not increase the spread of fungi by either the use of fungistats (notably malachite green) or removal of such eggs. Some farmers prefer to buy in eggs from another source rather than produce their own eggs. In this case it is necessary to wait until they are 'eyed eggs' (i.e. the embryonic eye has become visible) before moving them as then the eggs become far more resistant to the trauma of handling and transport. It is also extremely important to ensure that any bought-in eggs are bathed in a suitable disinfectant before introduction to the hatchery.

Many salmon farmers, in particular, now insist on the supply of eggs being certified free from certain diseases, and imported eggs into a country almost invariably have such stipulations. Nowadays, single fish testing has become the norm and tissues and gonad material from each potential brood fish are tested while the green eggs are incubating. Eyed eggs are then only released for sale following a negative test result.

After hatching has occurred the young fish feed on the contents of their yolk sac for several weeks and are called yolk-sac fry or alevins. The farmer must clean away the empty egg shells and ensure that water flows are adequate to supply oxygen and remove metabolic wastes. This latter task becomes increasingly critical once the fry have commenced to feed, at which stage they are usually transferred to some form of early rearing facility. Young rainbow trout cannot generally be reared in earth ponds due to the danger of whirling disease at this stage. However, in regions where this disease occurs, it can be avoided if the fry are kept in concrete or fibreglass tanks until they reach about 7 cm in length.

As the fish grow they require to be graded in order to prevent a large spread in unit weight among a particular stock, which would allow the larger fish to grow to the detriment of the smaller. Grading is normally performed once while within the fry tanks and then up to four times while the fish are within the on-growing facility. The method of grading depends on the size of the fish as well as the type of farm, and there is considerable scope for automation. Other important daily routines, such as cleaning inlet screens, filters and tanks, and repairing eroded banks and channels, are vital to prevent losses due to systems failures. Rainbow trout are usually harvested within the weight range 200–500g and the production cycle is determined by water temperatures, ranging from under 12 months to 18 months or more.

Atlantic salmon grow more slowly than rainbow trout during early freshwater life, but after transfer to the sea as smolts, growth rates often exceed those of trout, and salmon are also harvested at a much larger size. The traditional farming cycle for Atlantic salmon in Europe is outlined in Fig. 3.4. Originally the freshwater phase of salmon production usually took either about 18 months to produce 'one-year-old' smolts (S1) or less frequently 30 months to produce 'two-year-old' smolts (S2), these smolts then being transferred to sea from March to July but most commonly during May or June at approximately 25–75g.

A recent development in smolt production has been the development of out-of-season adaptation to saltwater phase by the process of 'jet-lagging' or altering the young fish's daylight pattern so that it undergoes saltwater adaptation either before or some time after the normal May period for smolting (November in the southern hemisphere). Such smolts are known as S0 or $S\frac{1}{2}$ smolts.

The seawater cycle takes a further 1–2 years depending on whether the salmon mature early as grilse the following year or carry on growing to be harvested during their second year at sea. Harvest

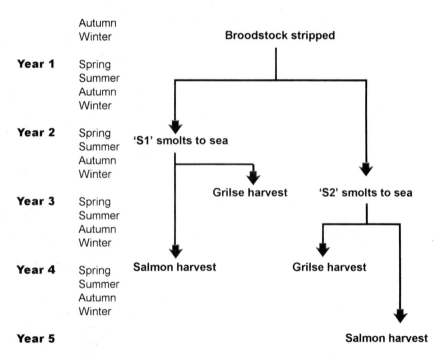

Fig. 3.4 Traditional farming cycle for Atlantic salmon in Europe.

weights for grilse and salmon vary enormously but are typically in the order of 2.5 kg and 4–5 kg respectively, after an overall production cycle from the egg ranging from about 2.5 years for S1 grilse to about 3.5 years for S2 salmon, although use of S$\frac{1}{2}$s will reduce this period accordingly. As with trout, a number of the better performing fish are selected for use as salmon broodstock, to be stripped of their eggs when mature in order to complete the cycle, although most farmers now depend on specialist egg producers for supplies from improved genetically selected broodstock.

3.4   FRESHWATER FARMING SYSTEMS

There is a wide variety of technological systems available to aid the three stages of incubation, early-rearing and on-growing. For incubation, traditional wooden hatchery trays have been replaced by vertically stacked incubators or vertical embryonators, which permit economies in space and water usage and ease operating problems. Early-rearing tanks are now rectangular or circular in design and are almost invariably made from fibreglass.

On-growing facilities comprise the major item of capital cost on trout farms and are often similar to the Danish systems of excavated earth ponds in parallel, feeding into a central outlet channel which also contains fish. These earth ponds are unlined and usually measure about 30 × 10 × 1.5 m deep. Such earth pond systems are often used where the terrain is suitable, easily excavated and not too porous. Each pond requires about three exchanges of water per day in order to hold over 1 t of fish at a stocking density of about 5 kg/m^3.

There has been a trend towards more intensive systems such as raceways, which were pioneered in the USA and usually comprise parallel sets of narrow channels in series. There are often falls from one channel to the next in the series with splash-boards for aeration purposes. Raceways are usually constructed in materials such as concrete, which resists erosion and permits higher flow rates than earth ponds, and correspondingly higher water exchange rates and stocking densities (e.g. three exchanges per hour and 32 kg/m^3, respectively).

Tanks, either square or circular, and made from fibreglass are increasingly becoming the norm for freshwater stages of salmon farming. They are usually constructed with a peripheral inlet pipe and central drain in order to provide a vortex and a consequent degree of self-cleaning. Unlike the cheaper earth ponds, raceways and circular tanks are quickly drained, cleaned and disinfected; also, the more intensive stocking and design of such systems permits better observation and eases operations such as grading and feeding, i.e. they give better control. Although more commonly used for marine culture of salmonids, floating cages are also used for rearing trout or salmon smolts in freshwater lakes and gravel pits. However, these systems are increasingly being replaced by flow through systems because of problems with their sustainability.

Freshwater is an increasingly scarce resource and suitable supplies can be evaluated for fish farming in terms of both their minimum flow rate and temperature range. These factors determine the stocks that may be held and the growth rates that may be achieved. For example, under average Scottish conditions, production of 1 t p.a. of trout to a unit weight of about 200 g generally requires in the region of 4.5–7.5 l/s (60–100 gpm) and an average production cycle of about 18 months can be assumed. At higher average temperatures available elsewhere, more oxygen, and hence water, is needed to produce 1 t of trout, but their growth will be faster.

Table 3.1 illustrates how the water requirement of rainbow trout varies with temperature, given a minimum dissolved oxygen level of 5.5 mg/l. To increase production levels where water is a limiting factor, it is possible to re-oxygenate the water using artificial aerators, assuming reasonable measures are taken to prevent any build-up of suspended solids. After dissolved oxygen and then suspended solids, the next limiting factors to water re-use are ammonia (particularly with alkaline water at pH values over 7) and increased carbon dioxide levels.

Systems now exist for recirculation of water with minimal make-up, which may be economical under certain circumstances. The capital and operating costs are high, however, since large volumes of biological filtration and complex and sophisticated pumping systems are necessary.

The choice of systems must take into account the relative problems and advantages of each in terms of the natural topography of the site. For instance, the use of pumped water will often permit construction of a farm where gravity-fed water is not available, e.g. beside a lake. Thus, the ultimate design of a farm is likely to reflect how far the farmer is prepared to go by way of maximizing production while keeping the inherent risks down to what he or she considers accept-

Table 3.1 Water requirement for 1 t of 200 g rainbow trout assuming 100% saturation of freshwater with oxygen

Water temperature (°C)	Oxygen consumption by 1 t of trout (kg/day)	Water requirement (l/s) (assuming dissolved oxygen of effluent = 5.5 mg/l)
6	2.6	4.3
8	3.4	6.2
10	4.3	8.6
12	5.1	11.2
14	6.0	14.3
16	6.8	17.7

able, taking account of site conditions. The possibility of major losses due to disease is one aspect of risk which the fish farmer can most readily reduce by good management.

3.5 MARINE FARMING SYSTEMS

Compared with freshwater, the marine environment has a number of advantages for rearing salmonids. There is generally much less competition for seawater, which is less likely to be polluted than surface freshwater. Also, seawater represents a very stable physical and chemical environment with high buffering capacity due to its dissolved salts. Whereas seawater contains less dissolved oxygen compared with fully saturated freshwater, this drawback is usually offset by the abundant availability of seawater with far less fluctuation in temperature, etc. These advantages have encouraged the on-growing of rainbow trout in seawater, although formerly trout were farmed entirely in freshwater. However, salmon require both freshwater and marine facilities for satisfactory growth through the entire life cycle.

The various alternative methods for rearing salmonids under marine conditions are summarized in Fig. 3.5. By far the commonest system involves the use of floating cages moored to the seabed. These usually comprise $1–2\,cm^2$ mesh nylon bag nets suspended from a square or circular flotation collar made of buoyant material such as polystyrene or rubber. A square cage has typical bagnet dimensions of $15 \times 15 \times 10\,m$ deep and weights are attached to the bottom of the net to prevent any distortion causing damage to the fish inside. Marine fouling of the nets must be checked by non-toxic antifouling paints or by appropriate net changing to avoid reducing the effective water exchange rate. Care must be taken to ensure that cage systems

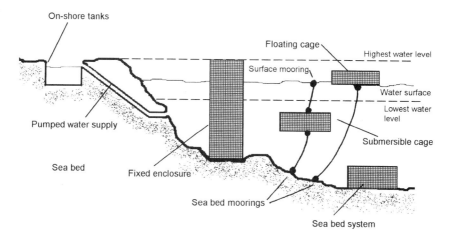

Fig. 3.5 Range of rearing options open to the marine farmer (after Milne, 1972).

are fitted with external predator nets both above and below the water level to avoid losses from birds and seals. Modern large scale marine farms are making increasing use of large barges, which incorporate cage groups with walkways, automatic feeding systems with feed silos, and staff accommodation; such farms are serviced by specialized work boats. The main risk is of storm damage to cages with torn nets leading to escaped fish, so even if the site is well sheltered, good cage design is of paramount importance.

Transferring salmonids from freshwater straight into full strength seawater (32–34 parts per thousand salinity) cannot be undertaken safely until Atlantic salmon have fully smolted or rainbow trout have reached about 100 g in size. Some smolt production units with access to pumped seawater start to introduce salmon parr to about 20% seawater towards the end of their first year, gradually increasing the proportion of seawater up to 100% by the following May. By this means it is possible to achieve a high proportion of S1 smolts from hatcheries with very cold freshwater temperatures, e.g. in Norway.

At the appropriate size, salmon and trout are transported by road, sea or helicopter from the hatchery to the marine farming site in specialized transport tanks. Provided salmon have smoltified and are in good condition, initial losses after smolt transfer are usually not more than 2%, although it may be 2–3 weeks before the fish are all feeding well. Salmon in particular should be handled as little as possible since loss of even a few scales will often lead to waterlogging and death at this stage. Stocking densities should be no more than 15 kg/m^3 and fish are either handfed two or three times a day or fed by automatic feeders linked to a calculated ration instead of to appetite. The grilse harvest will reduce stocking densities, but the remaining salmon grower population may still need to be thinned out by grading prior to harvest to avoid overstocking, with risks of a growth check or worse. Overall, the same basic husbandry principles apply both to marine and freshwater systems, with success linked closely to care, diligence and simplicity.

Chapter 4
Infectious Diseases

Some of the major losses in salmonid culture are due to factors such as poisons, incorrect nutrition, or systems failures. However, most losses are caused by disease processes involving living agents. It is important to know that all animals and plants consist of greater or lesser aggregates of cells, which are small self-contained 'factories' making essential components such as bone, skin or nerve. With the single exception of viruses, the living agents that cause disease in fish are also cellular in nature.

In this chapter diseases caused by living agents are grouped together according to the aetiology, i.e. the nature of the infectious agent involved. This is because many diseases can be caused by closely related organisms and rational treatment of each group of diseases is dependent on a knowledge of the differences between their agents.

These diseases are usually classified as:

* parasitic
* bacterial
* viral
* fungal.

4.1 PARASITIC DISEASES

A parasite is an animal that spends part or all of its life living at the expense of another animal, known as the host. Parasites may be found in every tissue of the host and are particularly common on external surfaces such as the gills or the skin. In the diseases that follow, the host is the salmonid fish but the parasites are of tremendous diversity. In the wild, all salmonids have small numbers of parasites, causing little if any injury, but in the farm or hatchery situation they can build up to considerable levels with harmful results. Parasites range in size from the microscopic to those that can be easily seen with the naked eye. They are usually divided up into the Protozoa, or single celled parasites, and the Metazoa, or multicellular parasites (Table 4.1).

Table 4.1 Classification of salmonid parasites of economic significance

Protozoans	(1)	*Ceratomyxa*
	(2)	*Costia* or *Ichthyobodo*
	(3)	*Henneguya*
	(4)	*Ichthyophthirius*
	(5)	*Hexamita* or *Octomitus*
	(6)	*Myxosporea*
	(7)	*Oodinium* and *Cryptocaryon*
	(8)	PKD agent
	(9)	*Plistophora* or *Loma*
	(10)	*Paramoeba*
	(11)	*Scyphidia* complex
	(12)	*Trichodina* complex
Metazoans		
Acanthocephalans	(1)	*Acanthocephalus* spp.
Cestodes	(1)	*Diphyllobothrium*
	(2)	*Eubothrium*
	(3)	*Triaenophorus*
Crustaceans	(1)	*Argulus*
	(2)	*Caligus* and *Lepeophtheirus*
	(3)	*Ceratothoa*
	(4)	*Lernaea*
	(5)	*Salmincola*
	(6)	Others (e.g. *Ergasilus*, *Achtheres*)
Nematodes	(1)	*Anisakis*
	(2)	*Cystidicola*
	(3)	*Eustrongylides*
	(4)	Others
Trematodes		
Monogenetic flukes	(1)	*Dactylogyrus*
	(2)	*Diplozoon*
	(3)	*Discocotyle*
	(4)	*Gyrodactylus*
Digenetic flukes	(1)	*Cotylurus/Apatemon*
	(2)	*Cryptocotyle*
	(3)	*Diplostomum*
Others	(1)	Lampreys
	(2)	Leeches
	(3)	Mussel glochidia

PROTOZOANS

Protozoa are the smallest parasites and they can only be seen properly with a microscope. Many can survive unfavourable conditions by having a spore stage in their life cycle, when their cell wall is very

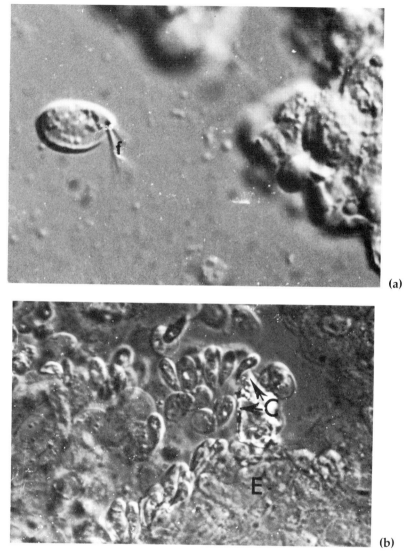

Fig. 4.1 *Costia necatrix* or *Ichthyobodo necator*. This parasite has a free swimming and an attached stage. It is small and here shown magnified at least 1200×. Thus when examining material under the microscope for the presence of *Costia*, the highest magnifications must be used. (a) *Costia* in its free-swimming form, showing the obvious pear shape and the flagella (f). (b) *Costia* feeding on the edge of a piece of salmon epithelium (C, *Costia*, E, epithelial cells). (c) Two attached forms of *Costia* (C) feeding on skin cells as observed in a wet smear preparation at high magnification (3000×). (d) *Costia* feeding on the edge of a gill lamella as seen in a typical wet smear preparation from a small salmon fry. (*Courtesy of Dr D. A. Robertson.*)

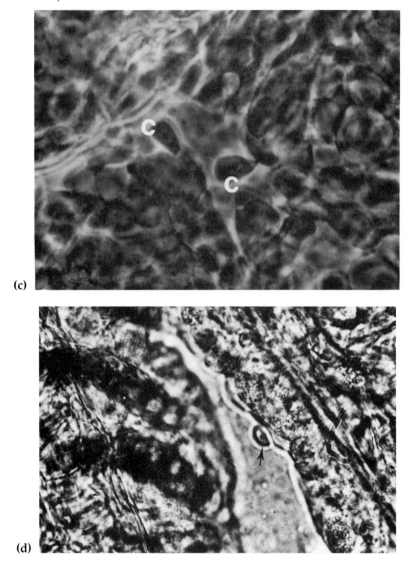

Fig. 4.1 *Continued*

resistant to heat, disinfectants and drugs. They are very varied in size and shape and live mainly on the skin and gills of salmonid fish, although a few serious diseases are caused by Protozoa that infect internal organs.

(1) *Ceratomyxa*
Ceratomyxa are myxosporean parasites like the causative agent of whirling disease. They can cause damage to almost any soft tissue where they are recognized by the characteristic shape of the spores.

(2) *Costia or Ichthyobodo*

Costia or Ichthyobodo is a very small pear-shaped protozoan (Fig. 4.1 and Plates 5a and 5b) which propels itself by means of whip-like hairs called flagellae. It is found on gills and skin surfaces and is of major importance in fry. *Costia* is about the same size as fish skin cells when observed in the microscope, but can be distinguished from them by its motility. The fish skin cells are non-motile, i.e. they do not move about. *Costia's* small size means that it is readily overlooked and it is best identified in the microscope by its jerky spiralling movements. It has now been found to affect salmonids in seawater as well as freshwater.

(3) *Henneguya*

The *Henneguya* group of parasites is found in the muscle and skin of wild salmon and sea trout, and is responsible for milky-flesh disease. It is characterized by tadpole-shaped spores, with two eye-spots or 'polar capsules' (Fig. 4.2).

(4) *Ichthyophthirius*

Ichthyophthirius is the causative agent of white spot or 'Ich'. It is a natural parasite of carp and goldfish and its occurrence in salmonids can usually be traced back to contact with coarse fish, although it is now endemic in most farmed salmonid waters and has occurred in Scottish cage farms in freshwater lochs with no coarse fish in them. It grows within the skin of fish and has a complex life cycle involving multiplication both on the host and in the water (see frontispiece (a)). The adult parasite, as it emerges from the fish, is a large (up to 1 mm), round, hairy, often brown-coloured parasite. It has a very obvious

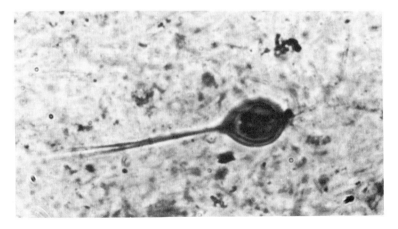

Fig. 4.2 *Henneguya*. This sporozoan parasite is shown here in a smear from the muscles of a coho salmon. The spores are encapsulated with two projections extending from the rear. This smear is magnified some 1200×. (*Courtesy of D. Bucke.*)

and characteristic horseshoe-shaped nucleus, and moves very slowly, if at all. It breaks out of the white spot on the fish, encysts on the bottom of a pond, and out of the cyst pass some 500 small pear-shaped infectious stages or tomites, which invade any fish with which they can make contact. They penetrate the skin and turn into the next stage, which is a rapidly revolving, globular, light-coloured parasite (about 0.2 mm). This develops into the mature adult parasite to continue the cycle at a rate which is increased at higher temperatures. Examination of a skin scraping or of material squeezed from a white spot frequently shows both of the host stages of the parasite (Fig. 4.3 and Plates 6a and 6b). Once an infection with *Ichthyophthirius* has run its course the fish are then immune to reinfection.

(5) *Hexamita or Octomitus*

Hexamita or *Octomitus* is a pear-shaped, very small and very active parasite of the gall bladder and intestine, usually of rainbow trout (Fig. 4.4). It moves rapidly by means of its long flagellae; in microscope preparations of smears of gut contents or gall bladder, it is usually the quick movement which attracts attention at low magnification. Its characteristic shape can be used for identification under the 40× lens.

The significance of *Hexamita* infection is not known with certainty. It seems to be mainly associated with fish that have other diseases, and is best considered as a disease of ill-thrift.

(6) *Myxosporea*

Myxosporeans are common parasites of salmonids in fresh and salt water and some species may cause losses. The myxosporea are traditionally considered as protozoa but in fact are a completely distinct group of animals. The stage normally seen in fish is the spore, which is thick-walled and contains two or more very distinctive structures, the so called polar capsules. In freshwater species, when spores are released from the fish they infest oligochaete worms living in the substrate and there undergo a second phase of development to produce another type of spore, the actinospore. These in time are released from the oligochaete worm and infest the fish. The life cycles of marine species are believed to be similar.

The best known myxosporean infecting salmonids is *Myxobolus* (*Myxosoma*) *cerebralis*, the causative agent of whirling disease in rainbow trout. If the parasite infects very young fry before the cartilage has ossified, it migrates through the gut wall to invade the skull, and severe skeletal deformities and behavioural disturbances will result. Although the parasite can only parasitize young fish before their cartilages have hardened into bone, once it penetrates the cartilages of the skull and spine it can, if present in any numbers, cause a severe reaction there. The damage that this causes affects the nerve function and the balance organ, resulting in the 'whirling' behaviour which gives this disease its name. Diagnosis is by chopping or scraping

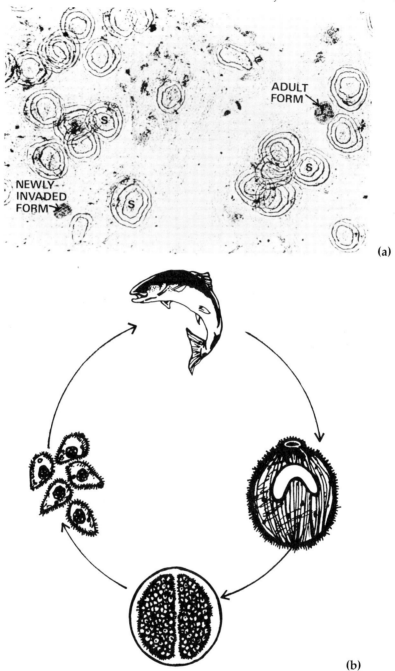

Fig. 4.3 *Ichthyophthirius.* (a) Smear of rainbow trout fry skin showing concentrically ringed scales (S) and the two stages of the Ich parasite. The small rapidly revolving protozoan has just invaded the skin, but the larger form is mature and ready to leave the host. The scales which are just under 1 mm in diameter give some indication of the relative sizes. (b) Life cycle of *Ichthyophthirius multifiliis.*

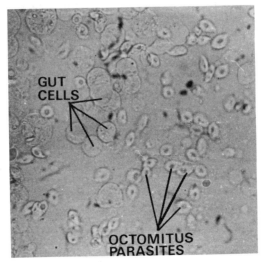

Fig. 4.4 *Hexamita (Octomitus)* and gut cells. These parasites are very active and difficult to see (or photograph). They are pear-shaped and half the size of the gut cells, which are also visible in this smear of gut contents viewed under 40× magnification (*Courtesy of Dr T. Håstein.*)

cartilage from the back of the head of suspect fish onto a slide and examining it for the characteristic parasitic spores (Plate 7).

Other myxosporean species which can cause problems in salmon culture include *Ceratomyxa* in Pacific salmon in North America and *Sphaerospora guttae* which can cause severe kidney pathology in salmon parr in Scotland. The parasite affects fish in the early summer and can cause swelling of the kidney and anaemia. Its biology and pathology are very similar to those of PKD (proliferative kidney disease) agent, but it is a distinct organism. *Kudoa* is a marine species that can cause marketing problems since it forms rather unsightly white cysts in the muscle. Control of myxosporeans is difficult since no chemotherapeutants are available.

(7) *Oodinium* and *Cryptocaryon*
The *Oodinium* and *Cryptocaryon* parasites have been observed in salmonids held at high density in marine aquaria, but have not yet emerged as major problems in the farm situation. *Oodinium* (Fig. 4.5) is probably a dinoflagellate but, like the protozoan *Cryptocaryon*, lives on the skin of salmonids in salt water.

(8) **PKD agent**
Although only described for the first time in 1974, in the first edition of this book, proliferative kidney disease (PKD) has now grown to assume the role of the most important parasitic disease of rainbow trout culture in the UK and many parts of Europe. It is also a serious condition in North America and probably has a world-wide distribu-

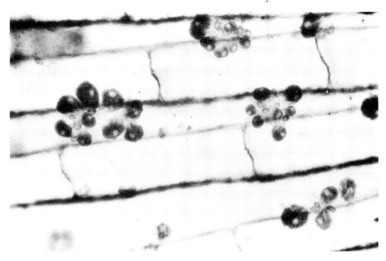

Fig. 4.5 *Oodinium.* These parasites are best seen on the fins or gills when they can be observed growing out from sites on fin rays. (*Courtesy of Dr R. Bootsma.*)

tion. The parasite, which has still not been fully described, is probably a myxosporean which fails to develop to the spore-forming stage. The parasite invades trout in the early summer and pathology is seen in August and September in the UK. The parasite is found in the interstitial tissue of the kidney and the spleen and causes severe anaemia. It is diagnosed in stained smear preparations (Appendix B) by its large round shape, pink cytoplasm and several nuclei (Plates 8a and b).

The life cycle of the PKD agent is not known, but it appears to affect fish in the first spring of their life, and there are some suggestions that it is a natural parasite of brown trout.

(9) *Plistophora* or *Loma*
Plistophora or *Loma* are the smallest parasites affecting salmonids and may occur widely, causing major losses, especially in North American trout farms. They have a complex life cycle, but the stage most readily recognized comprises a large number of spores which form a whitish cyst growing within the muscles or gills. These cysts may reach a considerable size, and the contents, when examined under 40× magnification, appear as minute, shiny spores, either oval or comma-shaped. The life cycle of the parasite is unknown but is believed to be direct.

(10) *Paramoeba*
Infections with *Paramoeba* in marine cultured salmonids can be a serious problem. The parasites are found on the gills and cause excess mucus production and the presence of white patches. Heavily infected fish are lethargic. The parasites are probably opportunistic

pathogens, and although particularly active at high temperatures, what it is that triggers their proliferation to the extent that they cause disease in fish is unknown.

(11) *Scyphidia* complex

The *Scyphidia* complex includes the organisms *Scyphidia*, *Epistylis* and *Glossatella*. All have the common feature that they merely use the skin of the fish as an attachment, or base, and feed off material passing them in the water (similar to barnacles living on a ship's hull). They are cylindrical or flask-shaped, and the stalk attaching them to the host may support several individuals. They are larger than *Costia* (see Fig. 4.1) or *Trichodina* and have a characteristic shape (Plate 9). If they are of any significance, they will be present in large numbers in skin scrapings. Their presence in significant numbers certainly indicates high levels of particulate organic matter in the water supply, which is normally an indication of less than ideal water quality.

(12) *Trichodina* complex

The *Trichodina* complex includes the organisms *Trichodina*, *Trichodinella* and *Chilodonella*. These parasites are larger than *Costia* and move slowly. *Trichodina* and *Trichodinella* are saucer-shaped and have sharp, rasping denticles or 'teeth' which damage the surface of the skin or gills when the parasite feeds. *Chilodonella* is a flat, heart-shaped organism. All have numerous bristly hairs called cilia. These parasites can be a severe problem on the skin and gills of fry or growers. They are identified in skin scrapings by their relatively large size (approximately five times that of *Costia*), slow movement and characteristic shapes as revealed in the frontispiece (b) and Plate 10.

METAZOANS

The non-protozoan parasites of salmonids are usually larger in size and may be members of any of a wide variety of zoological groups (see Table 4.1).

4.1.1   Acanthocephalans

The acanthocephalan or thorny-headed worms are very unusual worms. They are only found as parasites of fishes and have no equivalent in higher animals. They have, as the name suggests, considerable numbers of thorny hooks around the head and these are embedded into the lining of the gut. They can cause severe damage, with loss of weight if present in any numbers.

The larval stages occur within crustaceans, such as freshwater shrimps, or insects. Severe outbreaks have occurred in fisheries where large numbers of infected shrimps have been introduced in an effort to improve the fish colour or growth rate. The characteristic adult worms (Fig. 4.6) are readily seen with the naked eye in the intestine of affected fish, with the thorny head embedded in the intestinal wall.

4.1.2 Cestodes (tapeworms)

Tapeworms have complex life cycles involving more than one host. Although varying in length, each adult tapeworm has a head (scolex) with suckers or hooks with which it attaches to the host's intestine, and a series of segments (proglottids) which develop into egg sacs as the tape is extended from above. Tapeworms do not have mouths or intestines and obtain all their nourishment by diffusion of liquid food across their body wall.

As a proglottid, filled with eggs, reaches the end of the tape, it drops off and is passed in the faeces, releasing its eggs either within the gut or, more frequently, when it rots in the water. The life cycle is continued by the egg being taken up by an intermediate host, usually an insect or crustacean, where it undergoes a stage of development to become a procercoid; then the first host must be eaten by a second host, where a plerocercoid or infective form is produced. The plerocercoid's host is then eaten by the worm's final host, where the tapeworm matures into a new egg-producing tapeworm.

There are many variations on this standard fish – tapeworm life cycle. The salmonid may be an intermediate host, where the tapeworm encysts in the muscles or tissues as a plerocercoid, or it may be the final host containing the worm itself (Fig. 4.7).

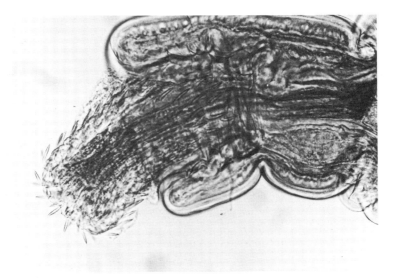

Fig. 4.6 *Acanthocephalus.* The typical thorny head is a rounded structure with arrays of briar-like 'thorns' which are used to attach the worm to the intestinal lining.

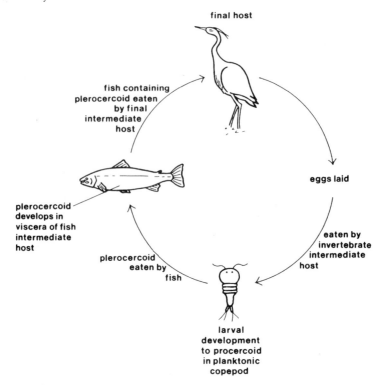

Fig. 4.7 Life cycle of a typical fish tapeworm.

(1) *Diphyllobothrium*

A number of species of *Diphyllobothrium* occur in salmonids world-wide, but in the UK two species, *D. dendriticum* and *D. ditremum*, are commonly found. Both species are found as plerocercoid larvae encysted in the viscera and more rarely the muscle of salmonids in freshwater. The adult worms are found in the intestine of fish-eating birds. Eggs passed out with the faeces of the bird hatch in water to release a free swimming larva which is eaten by the first intermediate host, a copepod of the genus *Cyclops*. The parasite develops within the body cavity of the copepod to a stage infectious to fish. If an infected copepod is eaten by a fish, the *Diphyllobothrium* penetrates through the gut wall and encysts within the viscera or sometimes the muscle. The life cycle is completed if an infected fish is eaten by a suitable bird. Because of the involvement of birds, fish and copepods in the life cycle, *Diphyllobothrium* is most abundant in fish reared in lakes.

The cysts of *Diphyllobothrium* are white and visible to the naked eye and they can represent a marketing problem if abundant. In young fish very heavy infections can cause mortalities, but usually they are more likely to have an adverse effect on growth rate and general condition of the fish. If sufficient cysts are present they may cause the viscera of the fish to become fused together.

Diphyllobothrium is killed by normal cooking and freezing processes. Cysts seem to disappear quickly if infected fish are moved to seawater. However, in freshwater they may persist for over one year.

(2) *Eubothrium*
Eubothrium crassum is a tapeworm occuring in both salmon and trout, in marine and freshwater. There are separate marine and freshwater races of the parasite. The worms are adult in the intestine of salmonids. Eggs shed from the fish intestine are eaten by the copepod intermediate host (in freshwater this is *Cyclops* spp.) in which the parasite develops to the plerocercoid stage infective to fish. The salmonid becomes infected by eating parasitized copepods. The involvement of a copepod in the life cycle means that fish reared in freshwater lakes are most heavily infected. The adult worms may be many centimetres in length and in heavy infections may form a cottonwool-like mass or even a thick 'rope' in the intestine (Plate 11). Although they may present something of an aesthetic problem, they do not appear to be very pathogenic, although it is suggested by some workers that the marine form may interfere with growth of the fish host.

(3) *Triaenophorus*
Triaenophorus is a problem in almost all waters where pike are plentiful. The pike is the final host, bearing the long white parasites in its intestine for much of the year but only releasing eggs in the spring. These hatch into infectious coracidia, which must be eaten by a crustacean within a couple of days. Here they develop into the procercoid and, if the crustacean is eaten by a salmonid (usually trout), they hatch out and migrate to the liver to form the second infectious stage, the plerocercoid (Plate 12). The final stage of the life cycle depends on the trout being eaten by another pike.

4.1.3   Crustaceans

The crustaceans, members of the same family as the more familiar lobster and shrimp, are among the most serious of all fish parasites. They are usually hard, flattened and clear or brown-coloured. They may be up to 1 cm in length and feed actively by inserting their proboscis into the tissues, or by rasping the surface.

(1) *Argulus* (fish louse)
Argulus is a flattened parasite, brown or even rather colourless, found moving over the skin and fins of fish in freshwater (Fig. 4.8). It can be a problem in trout culture in certain areas and can infect most species of fish. It is especially a problem in warm, stillwater fisheries. The parasite can readily move between fish and feeds on the body fluids of the host by means of a stylet which is inserted into the skin. *Argulus* can spread viral and bacterial pathogens between fish during

Fig. 4.8 *Argulus* (fish louse) is more commonly found on coarse fish and measures about 5 mm in length. It has a flattened body and a hard skin. It can readily swim from one host to another.

its feeding. The parasite leaves the fish to lay its eggs on suitable substrates such as stones or vegetation. The young parasites which hatch are immediately able to infect fish. It has a very hard coat which enables it to resist predation and can easily be seen with the naked eye on any part of the body surface.

(2) *Caligus* and *Lepeophtheirus* (salmon lice)

The salmon lice are parasitic copepods found on the skin of farmed and wild salmonids in seawater. They are flattened parasites, brown in colour and may be up to 1 cm in length (Fig. 4.9). The females bear distinctive paired egg strings which trail behind them. The parasite goes through up to ten stages in its life cycle. The first two stages are free living and act as a dispersal stage. The next stage is the infective copepodial stage which attaches to the fish and develops into the first of four chalimus stages. These are found especially around and on the fins and are attached by a frontal filament which is embedded in the skin. If present in sufficient numbers they may cause quite serious damage, especially to the dorsal fin. Most damage, however, is caused by the subsequent pre-adult and adult stages which are mobile over the surface of the fish and much larger. They browse over the skin and are usually most abundant on the dorsal surfaces, including the head. They can cause very serious damage, especially over the head where the skull can be exposed. Such heavily infected fish will be unmarketable, and will eventually die.

   Lepeophtheirus is specific to salmonids and is more dangerous than *Caligus* because of its greater size. *Caligus* occurs on a whole range of fish species.

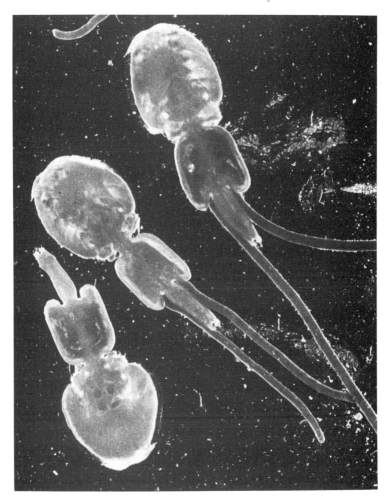

Fig. 4.9 *Lepeophtheirus*. The salmon louse is similar to the fish louse in size and shape but is dark brown in colour and the female has two distinctive light-coloured egg sacs trailing from the rear.

(3) *Ceratothoa*

Ceratothoa is an isopod crustacean causing serious effects on marine salmonids in Chilean waters. It locates in the mouth or branchial cavity, causing severe damage to the delicate gill tissue. Its full life cycle is not yet known.

(4) *Lernaea* (anchor worm)

Only the female of the *Lernaea* family is parasitic. When mature, it is a leathery, worm-like creature, attached to the muscle of the fish by an anchor-shaped head inserted to a depth of several millimetres. Eggs are released from the female, pass into the water, and undergo a number of maturation stages. Young females are fertilized in the

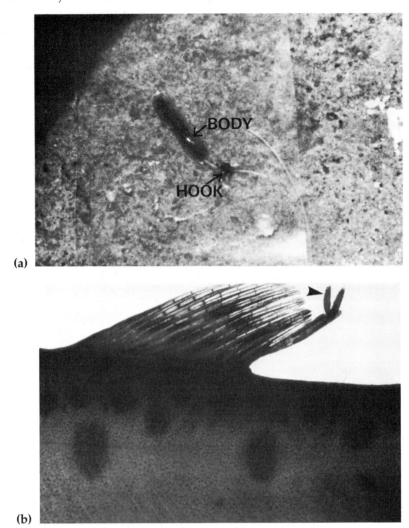

Fig. 4.10 *Lernaea.* Although they are closely related, the anchor worm is different in form from *Argulus*. (a) It has an anterior part consisting mainly of a rigid anchor which attaches to the flesh of the host. Its long fleshy tail protrudes from the fish surface. The parasite takes a long time to grow and it may be found at almost any size up to 1 cm. (b) The egg sacs (arrowed) are paired.

free-swimming stage and then become parasitic. This entails penetrating the skin of the fish, often at the vent, where the parasite grows its anchor and becomes readily visible (Fig. 4.10).

(5) *Salmincola* (gill maggot)
The gill maggot is commonly seen on wild Atlantic salmon and sea trout. It does not affect young fish due to its large size. Fresh-run fish

therefore are not infected, but once an adult fish has been in freshwater for any length of time, the parasite is usually present. It can
persist on the fish when it returns to sea, so that individuals that
survive to spawn again almost invariably have severe parasitic damage to their gills (Fig. 4.11). Other species of *Salmincola* have been
associated with gill damage in a variety of salmonid hosts.

(6) Others

A variety of other crustaceans (e.g. *Ergasilus*, *Achtheres*) have
been reported in salmonids. The gills are the most frequent site of
infection.

4.1.4 Nematodes (roundworms)

Roundworms are found in salmonids in many areas but rarely in
significant numbers. They usually have indirect life cycles, spending
their intermediate stages in insects or crustacea. Their main significance is in spoiling the table value of fish.

Fig. 4.11 *Salmincola*. The gill maggot is much lighter in colour than the
salmon louse and slightly smaller. It is sometimes found in considerable
numbers on the edge of the gill margins. (*Courtesy of W.M. Shearer.*)

(1) *Anisakis*

Anisakis is a marine nematode which uses wild salmon, among many other marine fish, as a second intermediate host. The first stage is spent in oceanic krill which are eaten by the migrating salmon, where it is usually found in small numbers on the surface of the liver or other viscera. Its final host is normally a porpoise, but it can invade the tissues of the seal or even man. It looks like a small, white, coiled watchspring on the surface of the abdominal organs (Fig. 4.12). Although of public health concern, it is only found in wild fish and never in farmed fish, which do not have an oceanic feeding stage.

(2) *Cystidicola*

Cystidicola are small (about 7 mm), white threadworms, frequently found in the swim bladder of salmonids. In certain fish, very large numbers may occur but, since they are not found in the muscles and do not usually cause harmful effects, they are of little significance.

(3) *Eustrongylides*

Eustrongylides are larval nematodes, red in colour, commonly found encysted in the viscera of wild trout in lakes. The adult worms are found in fish-eating birds. When the fish is opened, particularly if this occurs sometime after slaughter, the worms will be found crawling over the viscera (Plate 13). Although rather unattractive the parasites are killed by freezing or cooking.

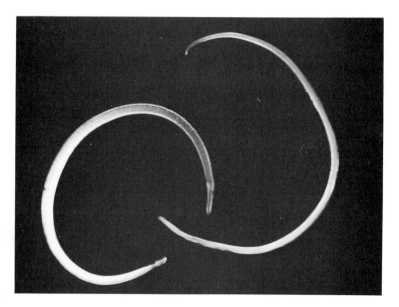

Fig. 4.12 *Anisakis*. These whitish-coloured nematode worms are found coiled on the surface of the viscera of marine salmonids. They are around 1 cm in length and may move actively when a fish is gutted. (*Courtesy of Dr K. MacKenzie.*)

(4) **Others**

Many other nematodes may occasionally be seen in the viscera of salmonids, but are generally of little importance.

4.1.5 Trematodes (flukes)

Flukes may be divided into two main classes: *monogenetic* and *digenetic*. Monogenetic flukes can spend their entire life cycle on one host, whereas digenetic flukes, like tapeworms, require more than one host. The adult flukes are visible to the naked eye.

Monogenetic flukes

(1) *Dactylogyrus*

Dactylogyrus is a common parasite of the gills of all salmonids. It destroys gill tissues by means of its suckers and hooks. It can multiply rapidly when there are damaged cells and tissue fluid for it to feed upon. It is up to 1 mm in length and, like other flukes, is best identified with a hand-lens or under the low power objective (Fig. 4.13).

Fig. 4.13 *Gyrodactylus*. The monogenean flukes *Gyrodactylus* and *Dactylogyrus* are usually about 0.5 mm in length and whitish coloured. Readily seen under the low power lens, they are distinguished by their rapid contractile movements. *Gyrodactylus* is viviparous, and in this photograph young parasites can be distinguished within some of the 'parent' parasites. (*Courtesy of Dr R. Bootsma.*)

(2) *Diplozoon*

Diplozoon is a parasite which consists of two flukes fused together. It is occasionally seen in gill preparations and readily identified, but it does not usually occur in sufficient numbers to cause disease.

(3) *Discocotyle*

Discocotyle is a gill fluke similar to *Diplozoon* in that it is larger and less common than *Dactylogyrus* and *Gyrodactylus*. It can be identified by its suckers and its lack of hooks, and is found on the gills of salmonids in freshwater. It is just visible to the naked eye and has a complicated arrangement of clamps at its posterior end which are used to attach to the gills. In wild fish *Discocotyle* is usually present in low numbers, but it can occasionally be present in sufficient numbers in pond culture to cause problems.

(4) *Gyrodactylus*

Gyrodactylus is the most important of the monogenetic flukes. It commonly occurs on the skin but occasionally affects the eyes or gills. It is a very small monogenean barely visible to the naked eye, but may occur in large numbers on the skin and fins of salmonids in fresh and saltwater. *Gyrodactylus* is viviparous, i.e. it gives birth to live young, and thus may increase in numbers very rapidly on a fish (Fig. 4.13). The parasite has a complicated arrangement of hooks at its posterior end for attachment to the fish. There are a number of species of *Gyrodactylus* on salmonids and most are not especially pathogenic. However, *G. salaris* has become identified as a potentially very important and deadly parasite of salmon in Norway. Originally found on salmon from the Baltic, *G. salaris* was apparently imported into Norway on smolts from Sweden and has decimated native salmon in some Norwegian rivers. The Norwegian salmon appear especially susceptible to the fluke, although in fish farms *G. salaris* is easily controlled using formalin. The parasite is not found in the UK and the Republic of Ireland or in the Americas although native salmon in the UK seem to be very susceptible experimentally. It is widespread in trout farms throughout western Europe.

Digenetic flukes

(1) *Cotylurus/Apatemon*

Cotylurus/Apatemon are found as larval or metacercarial stages encysted on the heart and viscera of salmon and trout in freshwater. They appear as small white cysts, often clumped at the tip of the ventricle. If present in sufficient numbers they may affect heart function, but they are not very serious pathogens. The final host of the parasites is a fish-eating bird.

(2) *Cryptocotyle*

Cryptocotyle is a very common parasite of marine fish and is occasionally seen on salmonids in the sea. The life cycle (Fig. 4.14) involves

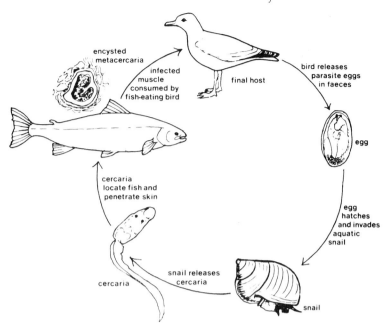

encysted
metacercaria

infected
muscle
consumed by
fish-eating bird

final host

bird releases
parasite eggs
in faeces

egg

egg
hatches
and invades
aquatic
snail

snail releases
cercaria

snail

cercaria

cercaria
locate fish and
penetrate skin

Fig. 4.14 *Cryptocotyle lingua.* The life cycle of this parasite depends on a multiplication stage within winkles (*Littorina littorea*), then release of the cercariae into the water to infect passing fish. The parasite invades the muscle of the fish to produce a black lesion and must be eaten by a sea bird to complete the life cycle.

marine birds, which infect whelks (marine 'snails'). In the summer very large numbers of cercariae (Fig. 4.15) are released from the whelks and actively seek and enter fish. Metacercariae are formed in the skin and the host fish lays down a readily visible black capsule around them, hence the name black-spot disease. The life cycle is completed when a bird, seal or occasionally human eats the fish and the adult fluke matures within the gut. The metacercaria in the fish is recognized as a small jet-black spot (about 1 mm) within the skin.

(3) *Diplostomum*

The metacercarial stage of *Diplostomum* occurs within the lens, humour and retina of the eyes of many species of fish in freshwater. *Diplostomum* are particularly attracted to the lens which may be completely destroyed. The adult parasite inhabits the intestine of birds and the eggs are passed into the water with the faeces of such fish-eating birds. There they hatch and the released larva invades gastropod molluscs of the genus *Lymnaea*. Enormous multiplication of the parasite occurs within the snail and in early summer large numbers of infective cercariae are released, which are able to penetrate into the fish usually through the skin around the lateral line, which may become reddened, and then migrate to the eyes. Invading cercariae

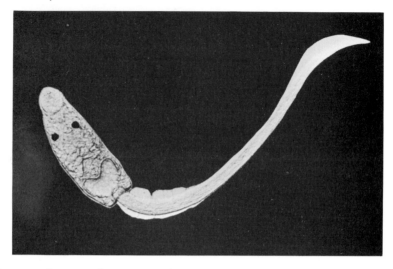

Fig. 4.15 *Cryptocotyle* cercaria. This is the infective stage for fish. It passes from winkles and, aided by its long tail, swims in search of a host. Once it attaches to its fish host, the tail breaks off. The cercaria is small (0.3 mm) and can only be seen in any detail at higher magnifications. (*Courtesy of Dr K. MacKenzie.*)

can cause irritation and flashing of the fish (twisting over on their side to give the appearance of a silvery shoal) if present in sufficient numbers. Fish normally become blind a few months after cercarial invasion, when the parasites have had time to grow and accumulate in the lens.

The life cycle of the fluke is completed when a fish-eating bird digests the eye and releases the metacercariae, which mature within its gut and produce many eggs (Fig. 4.16). Chopping up the lens of an infected fish and preparation of a wet mount allows identification of the metacercaria (Fig. 4.17) which can be seen flexing itself.

4.1.6   Others

There is a wide range of metazoan parasites which do not conveniently fall into any of the groups already considered. The three that are most important to salmonids are as follows:

(1) Lampreys
The lamprey is a very primitive type of fish which can be a serious problem to salmonids in certain areas, notably the great lakes of North America. It spends most of its life parasitic on other fish, but leaves the host to spawn in a redd, which is not unlike that of a salmon or trout. Lampreys have a large, round sucker at the mouth which allows them to clamp on to the skin of the host and then chew the tissues with their rasping teeth. Lampreys are easily recognized (Fig. 4.18) but show considerable variations in size, the freshwater

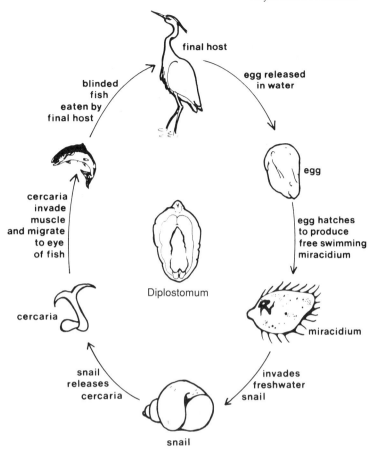

Fig. 4.16 *Diplostomum.* The life cycle again depends on a snail (*Lymnaea* spp.) and predatory birds. The parasite is the cause of eye fluke in salmonids.

lampreys being considerably smaller than the marine lamprey, which can be 50 cm in length.

(2) Leeches
Leeches are segmented worms which parasitize many animals, including man. They are also parasitic to salmonids and leave the host to breed in summer. They possess suckers at either end of their tube-like body and can loop around the surface of the host by using the suckers alternately. Leeches can swim freely (when they resemble elvers), but if feeding they attach to the skin of the host and suck its blood. They are easily recognized (Fig. 4.19) and measure up to about 5 cm in length.

(3) Mussel glochidia
All salmonids in freshwater are susceptible to infection by the larval form of the freshwater mussel (*Margaritifera*) (Plate 14). The freshwa-

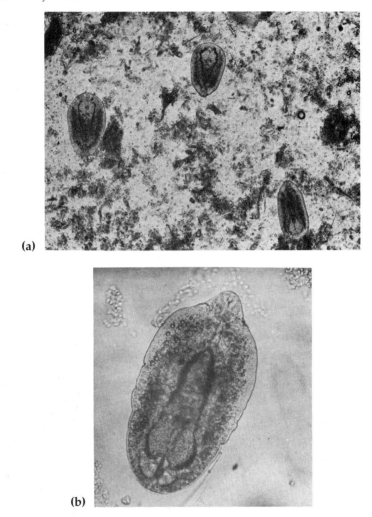

Fig. 4.17 (a) *Diplostomum* spp. This is a smear from the macerated lens of a blind rainbow trout. The metacercarial parasites are readily seen under the low magnification and their bodies can often be seen changing shape. (b) This individual *Diplostomum* shows the characteristic internal organelles. (*Courtesy of R.B. Stewart.*)

ter pearl mussel is a protected species in many countries. The larva is expelled by the mussel and must quickly invade the gills of a fish in order to develop further. This involves the parasitic larva biting off a piece of gill, on which it feeds as it attaches to underlying tissue. It is then walled-off by the host to form a cyst. These cysts can be identified as white specks on the gill lamellae (Plate 15), prior to the release of the next stage and completion of the mussel's life cycle. During the encystment stage the glochidia can allow secondary bacterial or protozoan infection to develop, and also, in warm water or at smolt

Fig. 4.18 Lampreys are very primitive fish which parasitize higher fish. This example of the river lamprey shows the typical round rasping mouth which grinds its way into the skin and attaches there to feed. (*Courtesy of W.M. Shearer.*)

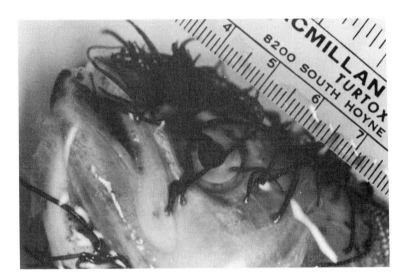

Fig. 4.19 Leeches are blood-sucking segmented worms. They may be 1 cm or more in length and swim with a looping movement. Heavy infections like this one are usually fatal. (*Courtesy of Dr Christina Sommerville.*)

transfer, the reduction in respiratory capacity which they induce can cause mortalities and even make a site unusable for smolt production over summer months.

4.2   BACTERIAL DISEASES

Bacteria are very small organisms which can only be seen with the high power lenses of the microscope. They abound in the environment and are capable of multiplying very rapidly. Many bacteria are of considerable value to man and are used, for example, to manufacture certain antibiotics and to produce yogurt. Some are, however, capable of growing in the host to its detriment. For example, in man, diseases such as tuberculosis, cholera and typhoid are caused by bacteria.

A number of species of bacteria cause disease in fish (Table 4.2). They vary greatly in the ease with which they can affect their host. For instance, certain strains of bacteria are capable of killing the host at almost any time, whereas many others can only cause harm at high temperatures, in overcrowded conditions, in very soft water, or in association with some other environmental circumstance which renders the host susceptible.

Some bacteria are normal inhabitants of fish skin or intestinal tract,

Table 4.2 Significant bacterial pathogens of salmonid fish species

Family	Genus	Species
Gram-negative species		
Cytophagaceae	*Flexibacter*	*Flexibacter columnaris*
		Flexibacter psychrophila
		Flexibacter marinus
Flavobacteriaceae		*Flavobacterium branchiophila*
Enterobacteriaceae	*Edwardsiella*	*Edwardsiella tarda*
		Yersinia ruckeri
Pseudomonadaceae	*Pseudomonas*	*Pseudomonas fluorescens*
Vibrionaceae	*Aeromonas*	*Aeromonas hydrophila*
		Aeromonas salmonicida var. achromogenes
		Aeromonas salmonicida var. salmonicida
	Vibrio	*Vibrio anguillarum*
		Vibrio ordali
		Vibrio salmonicida
Gram-positive species		
Bacillaceae	*Clostridium*	*Clostridium botulinum*
Corynebacteriaceae	*Renibacterium*	*Renibacterium salmoninarum*
Streptococcaceae	*Streptococcus*	*Streptococcus faecalis*
Acid-fast species		
Mycobacteriaceae	*Mycobacterium*	*Mycobacterium fortuitum*
		Mycobacterium marinum
Nocardia	*Nocardia*	*Nocardia asteroides*
		Nocardia kampachi
Rickettsias: obligate intracellular parasites		
Rickettsiales	*Piscirickettsia*	*Piscirickettsia salmonis*

Plate 14 The freshwater mussel *Margaritifera margaritifera*. This bivalve mollusc has a parasitic larval stage which encysts in the gills of fish. The adult mussel is embedded in gravel in the riverbed and releases its larvae into the water in very large numbers.

Plate 15 Glochidia infection. This rainbow trout has many small white cystic structures in the gill lamellae. These are the intermediate stages (glochidia) of the freshwater mussel. (Courtesy of Dr T. Hastein.)

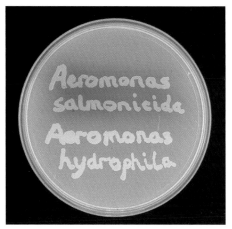

Plate 16 *Aeromonas salmonicida* producing typical dark diffusible pigment and *A. hydrophila* without pigment, growing on tryptone-soya-agar (TSA) plate. (Courtesy of Dr Val Inglis.)

Plate 17 Typical shallow ulcers of achromogenic *Aeromonas salmonicida* infection in the skin of a brook trout.

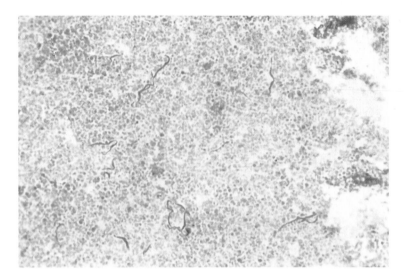

Plate 18 *Vibrio anguillarum* smear from culture stained by haematoxylin and eosin (Gram-negative).

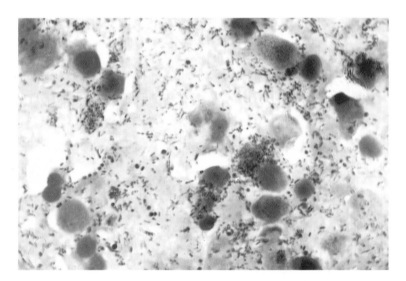

Plate 19 Smear from salmon kidney infected with *Renibacterium salmoninarum*. Stained by Gram's method. The bacteria appear as very small dark blue rods (Gram-positive).

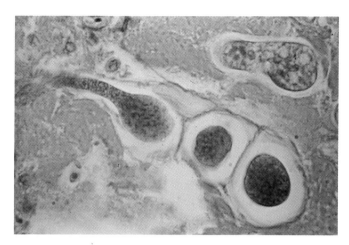

Plate 20 *Ichthyophonus*. The upper fungus seen in the smear is breaking out of its spore. Low power magnification is usually sufficient to distinguish *Ichthyophonus* (Courtesy of Dr J. S. Buchanan.)

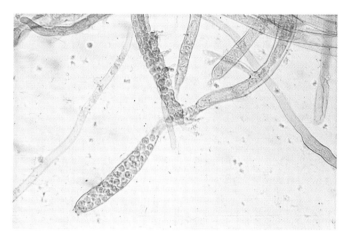

Plate 21 *Saprolegnia parasitica*. This fungus is one of the most serious of fish pathogens. Smears made from the cottonwool-like material, when examined under low power show the filaments and sporangia containing the motile zoospores which are released and spread the infection to new hosts.

Plate 22 Gas bubble disease. The site where the bubbles of gas (emboli) break out of capillaries varies with the age and species of fish. In these rainbow trout sac fry the bubble of gas has accumulated at the rear of the yolk-sac. (Courtesy C. H. Aldridge.) (Copyright, Unilever Research Laboratory.)

Plate 23 First feeding rainbow trout fry with *Phoma herbarum* infection showing the grossly distended abdomen, anterior to the yolk-sac, which is common in this disease.

Plate 24 (above) Unless fry infected with IPN die very quickly, they often show swelling of the abdomen and darkening of the back. (Courtesy of Professor N. O. Christensen.)

Plate 25 (above, right) Young Pacific salmon fry infected with IHN virus. They show the archetypal trailing faecal cast.

Plate 26 An example of the acute, haemorrhagic stage of VHS. There has been a haemorrhage into the testis and the gills show the pallor associated with consequent anaemia. (Courtesy of Professor N. O. Christensen.)

and when a fish dies or is killed they invade the body causing breakdown and rotting of the tissues. This putrefaction occurs very rapidly in fish carcases and is one of the reasons why 'unfixed' material is almost useless for laboratory diagnosis. These bacteria are rarely pathogenic in the living fish, but can complicate bacteriological examinations unless great care is taken.

Outside the laboratory, the only fish bacteria that can be readily seen under the microscope are certain large bacteria which may be identified in wet preparations of gills from diseased fish. However, it is usually necessary to isolate bacteria from fish tissues and grow them within the laboratory for purposes of identification. This can be done using plates of agar jelly which contains bacterial nutrients. Nowadays many hatchery biologists are expected to be able to culture bacteria and test their antibiotic sensitivity. Bacteria can also be stained so that they become easily visible under high power magnification in a microscope. Details of both bacterial culture techniques and staining methods are to be found in Appendix B.

4.2.1   Myxobacteria

Myxobacteria, often called the 'gliding' bacteria, comprise various micro-organisms which are normal inhabitants of water, mud and the body surface of fish. Certain species can cause specific diseases; however, in general, gliding bacteria are opportunist invaders of fish which have become vulnerable because of some stress, such as poor water quality or spawning.

Flexibacter columnaris
Flexibacter columnaris is the myxobacterium causing columnaris disease, which is so-called because the organism lines up in columns when placed on a microscope slide. It multiplies at high temperatures which is why outbreaks rarely occur below 25°C. It may be readily identified in wet smears at high magnification by the sliding sinuous movements of the organisms as they form into columns (Fig. 4.20).

Flexibacter psychrophila
The *Flexibacter psychrophila* group of myxobacteria are responsible for two conditions which affect salmonids at low temperatures – coldwater disease and peduncle disease. In addition, *F. psychrophila* (or possibly an ill-defined *flavobacterium* spp) has now been recognized as the cause of fry anaemia, also known as rainbow trout fry syndrome, a relatively new condition of cultured trout and salmon. The organisms are long and slender (Fig. 4.21).

Others
A variety of myxobacteria are responsible for disease in salmonids under stress. An important example is the colonization of damaged

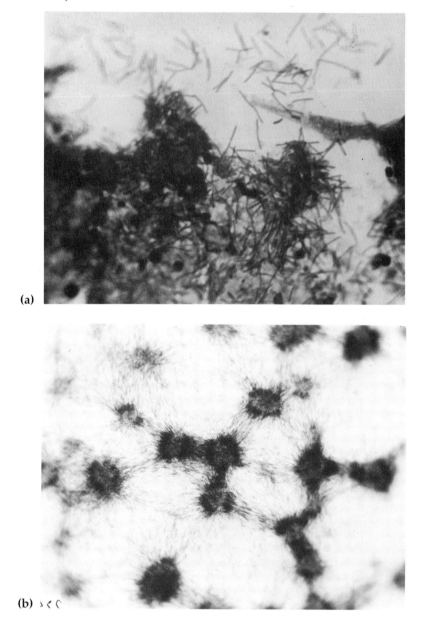

Fig. 4.20 *Flexibacter columnaris* on a wet smear. (a) The typical elongated bacteria. (b) Migration to form the columns which give the columnaris disease its name.

gills by *Flavobacterium branchiophila* in the condition known as bacterial gill disease. In this case, the bacteria may often be identified sticking tenaciously to the gills of affected fish in wet smears at 40 × magnification. They also play a prominent part in other diseases, especially fin rot and marine conditions such as eroded mouth and seawater precocity.

Fig. 4.21 Typical coldwater disease (*Flexibacter psychrophila*) organisms on wet smear.

4.2.2 Enterobacter

Enterobacteria are very common in freshwater, particularly if animal wastes are present. However, only two members of the entero-bacteria are associated with disease in salmonids. These are *Yersinia ruckeri* which causes a bacterial septicaemia, usually of farmed rainbow trout and occasionally salmon which is called enteric redmouth (ERM), and *Edwardsiella tarda*, the cause of occasional losses associated with septicaemia and foul-smelling enteritis in salmon or trout reared in heavily contaminated waters.

4.2.3 *Aeromonas* and *Pseudomonas*

The *Aeromonas* and *Pseudomonas* bacteria are responsible for a variety of diseases in wild and cultured salmonids. Although some strains are pathogenic at temperatures as low as 5°C, most do not multiply and produce disease below 10–12°C. They damage the host principally by means of their chemical poisons, or toxins, secreted during growth and multiplication within the host.

Aeromonas hydrophila and *Pseudomonas fluorescens*
Aeromonas hydrophila and *Pseudomonas fluorescens* are very common in freshwater, especially where there is a high level of organic matter,

such as sewage. They are particularly associated with haemorrhagic septicaemia in salmonids that have been stressed, e.g. at high temperature or spawning. Unlike *Aeromonas salmonicida* (see below) they can survive and multiply perfectly well in waters that do not contain fish, provided there are adequate levels of organic matter.

Aeromonas salmonicida

Aeromonas salmonicida is the cause of furunculosis, which is one of the most serious diseases of farmed salmonids. It differs from the other aeromonads in that it is an obligatory pathogen which does not normally multiply, or survive for any length of time, off the fish host, although recently some resistant forms have been found capable of survival over long periods in a dormant state. The period during which it can survive and remain infectious in the water depends greatly on whether there is a high organic load in the water, and particularly on whether the bacteria are contained within fish material, such as mucus, or are free. In pure water or in seawater, unassociated bacteria will last for only a few hours, but if the bacteria are associated with organic matter then they can remain viable for many days.

The normal type of *A. salmonicida* associated with disease in farmed salmonids produces a readily discernible diffusable brown pigment when growing in artificial culture which makes its identification straightforward (Plate 16). However, there are some types (the so-called achromogenic strains) which do not produce this, and although usually associated with diseases of carps they do occasionally affect salmonids, particularly the lake dwelling whitefish and chars. The condition of trout known as ulcer disease is now also known to be caused by achromogenic *A. salmonicida* (Plate 17).

4.2.4   Vibrio

Vibriosis is a disease of almost all species of fish in saltwater. The most important pathogen of the group is *Vibrio anguillarum* (Plate 18). This is a comma-shaped bacterium, and eels (*Anguilliformes*) are particularly susceptible, hence the name. Like aeromonads, vibrios damage the host by the action of their toxins. In human cholera, caused by an organism closely related to *V. anguillarum*, the toxin acts on the gut wall causing dysentery. *Vibrio anguillarum* probably multiplies mainly in the skin of salmonids, but the toxins produced act on the circulating blood cells, causing severe anaemia. This fact can be used to aid laboratory diagnosis using agar jelly containing blood as a nutrient. Growth of the organism, and hence disease outbreaks, occur more readily at temperatures above 11°C.

Vibrio ordali is a similar bacterium to *V. anguillarum*, but it has been particularly associated with Pacific salmonid culture, where it causes a disease characterized by focal muscle lesions rather than generalized haemorrhagic septicaemia. Another species, *V. salmonicida*, is specifically associated with heavy mortalities in Atlantic salmon at low temperatures in Norway (Hitra disease or coldwater vibriosis).

4.2.5 Renibacteria

Bacterial kidney disease (BKD) is caused by *Renibacterium salmoninarum* which can only multiply within the fish host, where it grows very slowly within the kidney, spleen and muscle tissue. The bacterium is very small and difficult to grow in culture because of its special nutritional requirements, so definite diagnosis can only be carried out in a specialized laboratory, although stained smears (Plate 19) often show the tiny rod-shaped bacteria in profusion.

4.2.6 *Mycobacterium* and *Nocardia*

Mycobacterium and *Nocardia* are responsible for chronic infections, such as tuberculosis and leprosy in man and other animals. They are characterized by a waxy outer coat which gives them great resistance to the host's defences and to disinfectants. *Mycobacterium fortuitum* and *M. marinum* are principally associated with fish tuberculosis (TB) and *N. asteroides* causes similar chronic problems.

4.2.7 *Piscirickettsia*

Rickettsias are small rod-shaped or coccoid Gram-negative bacteria which can only multiply within cells. *Piscirickettsia salmonis* causes salmonid rickettsial septicaemia. Initially diagnosed in farmed coho salmon in Chile, it is now recognized in northern Europe also, and is also found in rainbow trout and Atlantic salmon as well as Pacific salmon. The micro-organisms cannot be cultured on artificial media, but they do grow in salmonid cell lines. Diagnosis is generally based on clinical signs plus recognition of the rickettsia in Giemsa-stained smears.

4.3 VIRAL DISEASES

Viruses are the smallest known microbes. They are completely incapable of independent multiplication, unlike bacteria or fungi. They depend for their replication on being able to invade host cells and programme those cells into acting as their reproductive machine, so that the host is responsible for producing the virus for its own infection. Cells producing such a virus usually die.

Viruses can only be seen in an electron microscope which is capable of enlarging 50 000 times or more. If a full stop (.) was enlarged as much, it would be larger than a multistorey block. Virus diseases cannot be cured so it is essential that they are controlled by preventive medicine. There are five viruses that are definitely known to cause salmonid diseases, and other viruses may have significance in salmonid culture.

4.3.1   Infectious haematopoietic necrosis (IHN)

Infectious haematopoietic necrosis (IHN) is a viral disease of salmonids, which is currently confined to North America and Japan. Originally the group of diseases in wild and farmed Pacific salmon known as Oregon sockeye disease (OSD) and Sacramento River chinook disease (SRCD) was thought to differ from IHN of rainbow trout, but now all three are considered to be manifestations of the same disease complex, collectively referred to as IHN.

The virus can be transmitted by contact with infected fish, by feeding on infected carcases or by exposure to water from an infected source, and it is also carried on the surface of eggs from healthy, infected stock. It is particularly associated with Pacific salmon rivers where wild stock are endemically affected. The disease is not generally observed at temperatures exceeding 15°C, and, as with the other virus diseases, identification involves complicated laboratory procedures.

4.3.2   Infectious pancreatic necrosis (IPN)

Infectious pancreatic necrosis (IPN) was the first fish virus to be isolated. IPN is endemic in most parts of North America, Europe and Japan.

The virus can be carried via eggs, milt or faeces, and also exists in wild fish. Recent evidence also implicates mergansers and seagulls in transmitting it to distant watercourses. The virus damages the intestine and pancreas of affected fish (hence the name), and in healthy carrier fish may be regularly released within various secretions into the water to continue the cycle of infection. It occurs in both marine and freshwater stocks.

The virus is particularly resistant to adverse conditions. There are a number of different strains of IPN virus and they seem to have marked differences in pathogenicity. The most important strain occurring in Europe is called the Sp strain, while in the Americas VR 299 is the most significant. One strain known as the Ab strain is of low pathogenicity, but occurs widely, if infrequently, in wild fish; although not very important therefore in clinical terms, it has great significance in relation to diagnostic and certification procedures, since it is very difficult to distinguish from the other more pathogenic strains without extensive laboratory studies.

4.3.3   Infectious salmon anaemia (ISA)

Infectious salmon anaemia has only been definitely recorded in farmed Atlantic salmon in Norway, although outbreaks may have possibly occurred elsewhere. First recorded in 1984, it is characterized by very severe anaemia in its terminal stages and mortality is usually very high. Infection is spread via virus-infected blood or tissues, and until slaughter house wastes were prevented from being returned to the sea, these were a potent source of infection. It is gradually being eliminated from Norwegian stocks by careful hygiene and isolation and compulsory slaughter of infected stocks.

4.3.4   Pancreas disease (PD)

The virus agent causing pancreas disease (PD), a severe cause of mortality and chronic ill-thrift generally occurring in Atlantic salmon shortly after introduction to saltwater, has only recently been isolated. Little is known of its biology, but the severe acute degeneration of pancreas cells which it causes, and associated heart and muscle damage, lead to wasting, inappetence and often death in large numbers of affected fish. This sequel to PD is sometimes referred to as post-viral myopathy syndrome (PMS).

4.3.5   Salmon pox

This is probably a virus disease although the causative virus has not yet been isolated, and it affects the skin of wild and farmed salmon and occasionally rainbow trout, causing light coloured warts. These can occur during the freshwater or the saltwater stages of production, although in rainbow trout they have only been recorded in saltwater. Although unsightly, the warts eventually fall off and problems only arise if the resultant ulcers become infected with bacteria before healing.

4.3.6   Viral haemorrhagic septicaemia (VHS)

Viral haemorrhagic septicaemia (VHS) is a serious viral disease of farmed salmonids and occasionally marine fishes. It is also known as Egtved disease after the Danish village where it was first observed. The causative agent is a bullet-shaped virus similar to that of IHN, although VHS is confined to Europe and affects a different age group. VHS does not survive on the surface of eggs, which facilitates prevention and/or eradication of the disease. Although VHS is only of clinical significance to salmonids on the European mainland, it has occurred in marine fish in UK waters and also on the western seaboard of Europe. Its potential economic significance makes it extremely important in terms of international trade in eggs and fish and the necessary certification procedures.

4.3.7   Other viruses

A number of other viruses are associated with disease in salmonids. These include *Oncorhynchus masou* virus (OMV), a herpes virus of Japanese salmonids, which produces tumours around the mouth of affected fish. Viral erythrocytic necrosis (VEN) virus has been isolated from the blood of salmonids, but has not as yet been associated with any naturally occurring disease. It does, however, have some importance in relation to certification of fish for export to certain countries such as Australia.

Erythrocytic inclusion body syndrome (EIBS) virus is another virus affecting the red blood cells. It is a significant problem in Pacific salmon, especially in sea cage farms in Japan, but although the virus is widespread in Atlantic salmon stocks, it causes few problems.

4.4   FUNGAL AND ALGAL DISEASES

Fungi and algae are very primitive plants, of which only a few species are pathogenic to salmonids. Fungi do not contain chlorophyll, the green pigment of higher plants, and are therefore unable to use energy of the sun to manufacture food by photosynthesis. External fungal infections frequently follow damage to fish skin as most freshwater contains high levels of fungal spores, whereas internal fungal infections are usually associated with fungal contamination of fish feed. However, some fungi can produce toxins under suitable conditions of moisture and warmth. Thus, badly stored fish feed will allow ever-present fungal spores to manufacture these mycotoxins, which at low levels can cause reduced growth and at high levels can result in sudden large losses of fish. *Aspergillus* and *Penicillium* moulds can produce mycotoxins in fish feed and aflatoxins can produce liver tumours in trout, as described in Chapter 10.

Algae do not produce diseases in fish, but can be a potent cause of fish kills. This is because they constitute part of the freshwater and marine plankton, which can sometimes undergo a massive population explosion. Such marine algal blooms can turn whole seas red, orange, etc. depending on the species involved. This can kill fish due either to oxygen depletion or to nerve toxins released by the organisms, and this is described more fully in Chapter 11.

4.4.1   Black moulds

There is a variety of related mould species (e.g. *Exophiala*, *Phialophora*) which occur commonly in soil and vegetation and are known as black moulds. If the spores gain entrance into fish, the fungus may break out and invade the fish tissues. This usually occurs because of dietary contamination by fungal spores (compare with *Ichthyophonus* below), in which case the resulting fungal mass surrounds the gut region. However, some internal fungal infections occur around the swim-bladder and the fish may have picked up the spores directly from the surface film of water in hatcheries with a heavy atmospheric load of fungal spores. In any event, internal fungal infections are readily confirmed when the lesions are examined microscopically to reveal the characteristic branched structure of the fungal hyphae.

4.4.2 *Ichthyophonus*

The *Ichthyophonus* fungus is usually found in marine fish but is pathogenic to salmonids, even in freshwater, if they are fed with infected fish. If salmonids ingest the spore of *Ichthyophonus* the fungus breaks out and can invade all organs of the body. The organism is identified in lesion smears by its characteristic unsegmented hyphae and spore-containing sporangia (Fig. 4.22 and Plate 20).

4.4.3 *Saprolegnia*

The *Saprolegnia* group comprises a variety of closely related aquatic fungi which are consistently found in freshwater, particularly at low

Fig. 4.22 *Ichthyophonus.* This fungus grows throughout the tissues of infected fish. The fillet shows the whitish sago-like structures in the muscle containing the fungi. (*Courtesy of Dr J.S. Buchanan.*)

Fig. 4.23 *Scolecobasidium.* This soil fungus occasionally infects the muscle of brook trout and Pacific salmon. The fungus grows within swollen inflamed muscles, and the lesion can be seen on the side of the fish. (*Courtesy of W.T. Yasutake.*)

temperatures. They are usually associated with dead tissues and rapidly infect dead eggs of salmon and trout, from which they can spread to live eggs. Living fish are usually affected by invasion of *Saprolegnia* through lesions in the body surface and subsequent branching of the fungus over and through the tissues of the fish. The infection is spread by the release of motile spores from the spore container (sporangium) into the water. Identification is by the characteristic filaments and sporangia in wet smears (Plate 21).

4.4.4   *Scolecobasidium*

The *Scolecobasidium* fungus is normally found in soil, but it occasionally invades salmonid tissues causing hard raised swellings on the skin or in the kidney (Fig. 4.23). It is a branched fungus with small buds (conidia) containing the infective stages, which can be seen in wet mounts of the affected tissue.

Chapter 5
Disease Diagnosis on the Farm

Before fish farmers can hope to diagnose and control disease outbreaks, they must be able to appreciate when their stock is healthy and thriving. This is an art of husbandry which is only picked up by constant observation and experience. In addition it requires an understanding of how fish behaviour is modified by different conditions. Rainbow trout and brook trout feed much more readily than brown trout and salmon parr which are more shy. Water temperature, oxygen levels and health will all have a potent influence on the eagerness of fish to feed.

Daily inspection of the stock in each pond, tank, raceway or cage is an integral part of good husbandry. Before disturbing the fish in a pond, cage or tank the fish farmer should try to get a good look at them. Without the fish being aware of the farmer's presence, the farmer should seek to observe any unusual behaviour pattern and the distribution of the fish. Are they crowded around the inlets or outlets, or scattered around the sides of the pond or cage, or are they fairly evenly distributed as with healthy stock? Are they showing any unusual swimming behaviour, e.g. disorientated swimming, flashing (i.e. twisting over on their sides to give the appearance of a silvery shoal), scraping themselves on the edge of the pond, etc.? Having established this, it is good practice to approach the fish and watch their reaction when they first become aware of human presence. Do they 'start' suddenly as if excessively nervous, or do they merely swim more urgently, displaying a keen appetite? Cages at sea are more difficult to observe, but fish can be assessed at feeding in all except stormiest conditions.

Whatever method of feeding is employed, it is essential that wherever possible the fish be fed by hand at least once a day, preferably first thing in the morning. Eagerness to feed is a most important sign of health and should be assessed carefully. Certain signs of disease are often particularly evident at feeding, e.g. the frantic tail-chasing of trout with whirling disease, or the lack of balance in fish with swimbladder or balance organ obstructions. Any obvious signs of sickness should be looked for at this time, especially the presence of gasping fish, dark-coloured or discoloured fish, bulging eyes ('popeye'), eroded tails and fins, and, particularly in salmon cages, the black, thin 'poor-doers' that hang around the edge of the cage. This part of the inspection routine may often be made easier by the use of polaroid glasses which permit a better view of the fish.

A cardinal rule which cannot be overemphasized is that the tank, pond or cage should be examined for dead or dying fish, which should be removed, counted and disposed of daily. Dead fish are the most significant source of infection for other fish, especially if left to rot on the bottom. The disposal of mortalities is a matter of importance both from a fish-health point of view and also human aesthetics. Principally, however, disposal must be in such a way as to preclude the possibility of infected material entering waterways or the farm. This will usually involve burial with lime, but other methods are also used.

Occurrence of mortalities following grading is often evidence of poor health, which reduces the ability of the stock to cope with the stress of handling. If clinical evidence of disease is found, a variety of fish should be inspected, particularly if mortalities appear to be excessive. The vigour and conformation of the individual fish should be noted, as well as certain external characteristics:

- The nature of the mucous coating of the skin: is there excessive mucus and is there any discolouration?
- The opercula over the gills: are they shortened and, if so, do the gills appear normally coloured (there should be no strands of mucus or fungus visible)?
- The fins and tail: are they eroded with a ragged edge, or are they intact?
- Any other external signs: is there any evidence of a rash or white spots on the skin? Are there any sores, ulcers (i.e. holes), lumps or parasitic animals on the skin? Is the eye normal or does it show popeye, and is the lens clear or cloudy?
- In the case of smolts, has there been excessive loss of scales or darkening or other sign of failure to adapt to seawater?

If the cause of a particular problem is not immediately obvious, a post-mortem examination may be necessary. For this purpose, live fish which show frank evidence of disease should be taken and killed.

Fish that are already dead when provided for examination cannot be used for post-mortem examination as degeneration changes take place in fish tissues immediately after death, even where the temperature is very low. Bacteria also invade the tissues, both from the water and from the intestine, as soon as death occurs, and so such dead fish are useless for microbiological analysis since false positive results will invariably occur. The post-mortem should be undertaken well away from any live fish, and care should be taken to ensure that it is performed hygienically. Thus, after the examination is completed, any fish remains should be properly and safely disposed of, and the fish farmer should adequately disinfect him or herself and the utensils.

The following equipment is required for a satisfactory fish post-mortem: microscope (see Appendix A), microscope slides and coverslips, scalpel, mounted needle, fine scissors, hand knife, hand lens, and bottles of 10% formal saline. In addition, in many farms bacterial culture and sensitivity testing is now carried out and further requirements for this are indicated in Appendix B. Fish should be killed either by a blow on the skull or by decapitation, and this should be done carefully to avoid removing mucus and possible skin parasites.

5.1   PROCEDURE FOR EXTERNAL EXAMINATION

Scrape a scalpel blade across the skin surface and place the collected mucus on a glass slide. Add a drop of water and tease out the mucus with a mounted needle. Place a coverslip over the mucus; this is a wet preparation of a skin scraping for examination under the microscope. Remove the gills, or a single gill-arch, with the scalpel or fine scissors (Fig. 5.1) and place onto a slide as before to give a wet preparation of gills. Slit the eye either *in situ* or after dissecting it out of its socket. Remove the lens and associated fluid with a mounted needle. Chop the lens with a scalpel and examine a wet preparation at low power (eye flukes present may also be seen with a hand lens). Any unusual lesions which are visible on the outside of the fish should be dissected free and placed in 10% formal saline if more detailed laboratory examination is required.

5.2   PROCEDURE FOR INTERNAL EXAMINATION

Slit the fish open along the midline of its belly from between the gills to the anus. Look carefully at the inside of the abdominal cavity and its contents. Make particular note of the colour of the various organs, the presence (and colour) of any fluid present, and also any swellings or unusual lesions, e.g. haemorrhages. Remove the gut and slit along its length with the fine scissors. Smear the contents of the gut onto a slide and make a wet preparation in order to search for any gut parasites, e.g. *Hexamita*. If any unusual lesions are seen with the naked eye, these should also be placed in 10% formal saline for more detailed laboratory examination. Under certain circumstances (e.g. after a fish kill or if certain virus diseases are suspected), small blocks of liver, kidney, gut (at the pyloric caeca), heart, spleen, gill and lateral line area of skin and muscle should be removed and placed in formal saline. The formal saline preserves the tissue until it can be sent to the diagnostic laboratory, but because formalin does not penetrate tissues very well, it is important to ensure that blocks of tissue are no more than 0.5 cm in size and that large volumes of fixative (approximately 25 ml per block) are used.

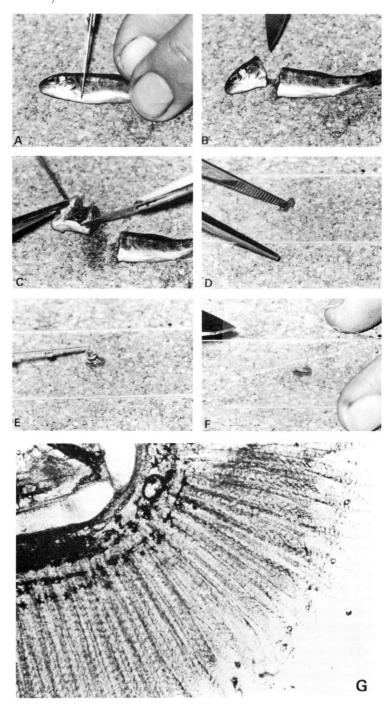

If microbiological sampling is to be undertaken because of the suspicion of a viral or a bacterial infection, some of the fish should be deep frozen immediately after sacrifice, or preferably forwarded to the diagnostic laboratory alive. Formalin fixation destroys microorganisms, and so formalin fixed material is useless for microbiological examination.

It is very important that specimens fixed and suspended in formalin, or frozen, are not forwarded to the diagnostic laboratory via the public postal service. Not only is this dangerous, but it is also illegal under the rules of the International Postal Union. If formalin fixed material is to be forwarded by post then it must be removed from the fixative and wrapped in damp cottonwool prior to sealing in a polythene bag and posted in a stout envelope.

Diagnostic laboratories vary in their capability and in their availability from country to country, and in most cases they are commercial organizations which have to charge a fee for their services. Wherever possible it is important to advise the laboratory by telephone or fax prior to forwarding samples, so that appropriate arrangements can be made.

Some farms are now carrying out their own preliminary bacterial screening and antibiotic sensitivity testing (see Appendix B). If such testing is to be performed, it is essential that proper aseptic technique is carried out at post-mortem. Following examination of the external surfaces of the fish, its abdomen should be gently swabbed with disinfectant and opened with a heat sterilized scalpel. Great care is required at this stage to ensure that only fish that have just been sacrificed are used and that external bacterial contaminants are not introduced from the hands or the fish surface.

Normally the bacterial conditions amenable to on-farm bacteriological examination will be generalized infections such as furunculosis or vibriosis. In such cases, the kidney tissue usually provides the best source of infected material, but the spleen and the heart blood may also be of value. In sampling the kidney, the swimbladder should be gently pushed aside, without touching the

---

Fig. 5.1 Preparation of a smear for a wet mount. Since protozoa may die rapidly after death of the host, it is important that all of these procedures are carried out promptly. **A** The fish is handled very carefully to avoid damage. **B** It is decapitated. **C** The material for examination is carefully removed (in this case the gill, but the process is the same for smears of skin or gut contents). **D** The gill filament is placed on a clean microscope slide (a larger gill would be dissected or teased out prior to examination). **E** A drop of clean saline (or unchlorinated water) is added. **F** A coverslip is placed over the preparation, with great care being taken not to include bubbles of air. **G** The preparation is then examined under the microscope. Always examine under low power first, and look for large parasites or the fast movements of smaller ones. This gill preparation is perfectly normal.

kidney surface, and a sterilized bacteriological inoculating loop, or better still a sterile inoculating needle, pushed into the kidney substance to remove a small amount of material for inoculation onto a suitable bacteriological isolation medium.

Where skin lesions are to be sampled, the surface should be seared with a hot knife to kill any surface contaminant bacteria before a sample is taken from just below the surface, with a sterile needle.

If isolation media are not immediately available, then samples may be taken using sterile bacteriological swabs. These should be kept cool and used within 24 hours for inoculation. Bacterial transport media are sometimes recommended, but in view of the risk of contaminants growing rapidly in such media, they are unsuitable for fish.

Fish farmers must always remain aware that in most countries there are specific, designated diseases, usually referred to as notifiable diseases, which must, by law, be reported to an appropriate governmental authority if their presence is suspected. Such regulations are essential for the well-being of the entire fish farming community as well as the wild salmonid stocks of our waters.

In many farms where eggs or live fish for export or for ongrowing elsewhere are produced, detailed regular certification by an approved laboratory has to be carried out. In this case the sampling of the required proportion of the stock for testing should be random, and is generally undertaken by the veterinarian responsible for the testing.

Chapter 6
Diseases of Eggs and Sac Fry (Alevins)

In the following chapters the important diseases of salmon and trout are described according to the stage in the growth cycle at which they occur. Diseases are usually closely related to the physiological state of the fish concerned and the particular husbandry, so diagnosis of a problem is made easier if the clinical findings can be considered in the context of the range of conditions that are known to occur at this particular time.

Two particular types of problem are discussed separately – fish kills and nutritional diseases. A fish kill (Chapter 11) may be defined as a sudden mass mortality among previously healthy fish which occurs within a period of 24 hours. The only infectious disease that can be confused with a fish kill is peracute furunculosis in young salmon. Nutritional diseases (Chapter 10), on the other hand, can be readily confused with many chronic infectious diseases. They generally develop over a considerable period, and may be manifested in a variety of different ways. With modern compounded diets, nutritional diseases are generally uncommon and a diagnosis of nutritional disease should normally be considered only after other more likely causes have been eliminated.

6.1 DISEASES OF EGGS

There appears to be little correlation between the size and colour of individual salmonid eggs and their viability. A certain number of infertile or 'blank' eggs is normally produced even by healthy broodstock, especially if they are stripped outside the optimum period. If more than 20% of a batch is infertile then it may be advisable to discard the entire batch.

Apart from eggs that are infertile, a proportion of live eggs will die during development, particularly if they are roughly handled as green eggs before reaching the 'eyed ova' stage. Such eggs, when they die, will inevitably become infected with fungus which then spreads to adjacent healthy eggs (Fig. 6.1). Three different ways of controlling this problem are usually practised. Firstly, it is now normal practice to stress green eggs by 'shocking' them in order to separate those that are going to succumb from the rest, thus allowing all dead eggs to be removed at once. Thereafter, daily 'picking' of dead eggs can be carried out with a pipette, or alternatively the

Fig. 6.1 Dead salmonid eggs. This photograph shows light coloured opaque eggs which have died during incubation. These readily become infected with *Saprolegnia* or other pathogens which can spread to healthy eggs such as the normal (eyed) eggs also present.

hatchery manager may prefer not to pick dead eggs but to prevent fungal spread by daily disinfection with malachite green or formalin before discarding the dead eggs automatically using an egg counter shortly before hatching. Malachite green is, however, a chemical which may have some slight potential for human health hazard so in some countries its use is not allowed.

Under certain circumstances eggs may show disorders, notably the occurrence of white or coagulated yolk spots inside the egg (and later in the yolk-sac of fry). This is thought to be due to the effects of heavy metals in the water, notably zinc and copper. It is essential therefore that pipes that are galvanized or made of copper are excluded from the hatchery. Occasionally eggs will develop a soft and sticky consistency and tend to clump together. Although the cause of these problems is not fully understood, it is thought to be associated with excess ammonia in the water. In this case the best course of action is to increase the water flow through the hatchery and thus flush the ammonia away from the eggs.

It should always be remembered that the quality of the water supply, while extremely important throughout a fish farm, is particularly so as regards the hatchery supply. The water temperature should not exceed 13°C. The water should be free of suspended solids, which will coat and choke the eggs. Dissolved iron salts, e.g. from iron pipes or natural deposits, will also precipitate onto the surface of eggs and cause similar losses. If suspended matter is a problem, for example after heavy spates, it may be advisable to incorporate a gravel filter bed at the inlet.

A particular problem in hatcheries in certain areas with soft water, such as Scotland, Scandinavia and Canada, is the occurrence of low pH spates with concomitant high levels of manganese or aluminium, resulting from so-called acid rain. This problem is discussed in Chapter 11. It has implications for hatcheries in that salmonid eggs will not hatch out when the hatchery water has a low pH and high levels of dissolved metal ions. Thus, hatching can be completely prevented if there are prolonged snowmelt or spate conditions in an acid rain area. Dramatic synchronous hatching may occur if the pH is chemically buffered with lime at such times.

Many infectious diseases can be transmitted via the eggs from one generation of fish to the next. Most pathogens released with the eggs and reproductive fluids at spawning from carrier brood fish are only able to attach to the surface of the egg, and with suitable disinfection (see Chapter 13) can be destroyed. Certain pathogens, however, are able to maintain themselves *inside* the egg, without killing it, and are released into the water to infect other fish at hatching or first feeding. Most important of these vertically transmitted pathogens are IPN virus and *Renibacterium salmoninarum*, which causes BKD.

Producers of high quality eggs or fingerlings generally have their hatcheries certified as free from these conditions, following regular examination and periodic monitoring of broodstock. The most satisfactory form of such monitoring is the system known as individual fish testing which requires sacrifice of each fish used as broodstock, with their fertilized eggs kept separately while detailed testing of the parents is carried out.

One of the best indicators of egg health is the quality of the ovarian fluids which are shed with the eggs. This should be completely clear, if somewhat viscid. If these fluids are markedly haemorrhagic, or cloudy, and particularly if these fluids are white, stringy or containing particulate matter, the eggs are likely to be contaminated and should not be used.

6.2   DISEASES OF SAC FRY (ALEVINS)

Hygienic hatchery conditions, important before hatching occurs, become crucial after this stage. Egg shells in particular will serve as a focus for the multiplication of fungi and parasites if they are not regularly removed. Attention to sanitary procedures will provide a better environment for the sac fry, which will thus have a better start in life.

Very small numbers of abnormally structured fry are not unusual. These 'monsters' may take the form of Siamese twins or the like (Fig. 6.2), and if their incidence is more than 1%, then that particular batch of eggs should be regarded as suspect. Apart from such occurrences, there are two abnormalities of the yolk-sac which are sometimes seen at this stage: blue-sac disease and deformed yolk-sac.

6.2.1   Blue-sac disease

If the yolk-sac increases in size so that the fry cannot swim in their normal position, then blue-sac disease may be suspected. The condition is brought about by an increase in fluid in the yolk-sac which takes on a blue-grey coloration (see frontispiece (c)). The underlying cause is probably the accumulation of metabolic products, and the problem is alleviated to some extent by increasing the water flows.

6.2.2   Deformed yolk-sac

Sometimes it may be observed that a fat globule within the yolk-sac has become herniated or pinched-off, so that the yolk-sac itself takes on a dumbbell appearance (Fig. 6.3). This condition often occurs in salmon where the yolk-sac fry have to settle on a smooth base of the hatching tray. It is alleviated by provision of suitable rough substrate such as grooves in the tray or gravel. In trout it is usually associated with inadequate water flows or high incubation temperatures, and

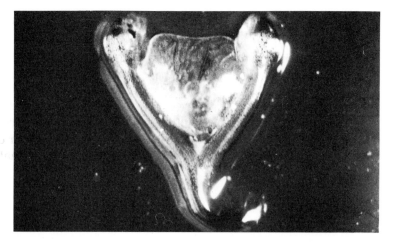

Fig. 6.2 'Siamese twins'. This is one of a number of relatively common congenital anomalies. The two fry are fused at the yolk-sac. (*Courtesy of Dr T. Håstein.*)

Fig. 6.3 Deformed yolk-sac. A rainbow trout showing a dumbbell-shaped yolk-sac, a relatively common anomaly associated with build-up of metabolic wastes in the water. (*Courtesy of Dr. T. Håstein.*)

increasing water flows or reducing stocking density may be the best prevention.

Where there is a disease problem of sac fry with no evidence of yolk-sac deformity, and if both the oxygen levels and flow rate of the water supply are adequate, then the possibility of gas embolism or parasites should be considered.

6.2.3 Gas-bubble disease

In yolk-sac fry, supersaturation of the hatchery water supply with dissolved gases leads to accumulation of bubbles in the yolk sac (Plate 22) and losses can take on the significance of a fish kill (Chapter 11). However, in less severe cases the bubbles of gas may not be obvious, and in slightly older fry it is the gills that are usually affected. Fry may swim in a disturbed fashion due to buoyancy alteration. The condition is similar to the 'bends' which may occur when human divers surface too rapidly following a deep dive. Affected fry frequently swim upside down or hang in the water, and the scale of losses varies considerably. This condition should be strongly suspected if, when the fish farmer's hand is placed in the water, a silver sheen of bubbles immediately forms on the surface of the skin.

When affected fish are examined under a lens, bubbles of gas may be observed in the skin or within the capillaries of the gill (Fig. 6.4). Although mortality levels in a stock of fry may be low, survivors of an outbreak rarely do as well thereafter as might be expected. The remedy is to blow off excess gas by aeration and to repair any leaks in pipelines or pumps which may be introducing air into the system. Where particularly cold water supplies enter a warm hatchery,

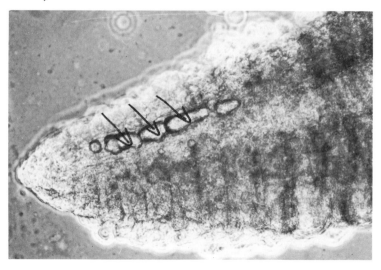

Fig. 6.4 Gas-bubble disease. The site where the bubbles of gas (emboli) break out of capillaries varies with the age and species of fish. In young Atlantic salmon the gill is the commonest site of gas-bubble formation. This preparation shows the tip of the primary gill lamella of a salmon fry. The bubbles (arrowed) are readily seen in such preparations. (*Courtesy of C.H. Aldridge.*) (*Copyright, Unilever Research Laboratory.*)

supersaturation can also occur causing gas-bubble disease in fish if there is inadequate mixing of the supersaturated water before it enters the hatchery tanks.

6.2.4 Costiasis

If daily losses of fry are occurring with no obvious external signs, then parasitic infection is one of the most likely causes. *Costia Ichthyobodo* may infect yolk-sac fry and, in order to confirm this, wet preparations of skin and gills should be examined under high power magnification. The problem is readily treated with formalin, but great care is necessary to ensure accuracy of dosing, as yolk-sac fry are very sensitive. Often costiasis occurs in association with a flush of spate water with high levels of suspended solids present and, if this is a regular occurrence, some form of filter should be incorporated in the system.

6.2.5 Fungal infection

First feeding fry feed very keenly as the yolk is used up and will snap voraciously at anything of the appropriate particle size. This can include fungal spores, especially where soil or dirt are associated with the water supply to the fry. The species of fungus involved is usually *Phoma* or *Saprolegnia*, and ingestion can result in generalized growth of the fungus, particularly in any remaining yolk. Affected fish have white rounded yolk-sacs, or if older are dark and show slight abdominal distension (Plate 23). High mortality may occur and there is no treatment possible. Insertion of a filter on the hatchery water supply and strict tank hygiene will help prevent the condition.

Chapter 7
Diseases of Early Feeding

First feeding of fry is a crucial stage in any hatchery operation. If undertaken too early, it will encourage infections of the yolk-sac as well as increasing the amount of debris in the hatchery, with all the concomitant problems to which a dirty environment predisposes. On the other hand, young fry that do not get onto the feed sufficiently quickly can live for a considerable period on the food reserves of their yolk and then their body fat. These fish may not seem to be failing, but they become very slender with large heads, i.e. 'pinhead' fish (Fig. 7.1), and then may start to die in considerable numbers. This situation is particularly common when first feeding of Atlantic salmon has been retarded and the fish have gone beyond the point of no return. No matter how much food is available, such fish cannot then make good the early starvation. Mortality of this nature can be quite sudden in onset and resemble the effects of IPN virus or IHN virus infection, although the pinhead appearance of the fish should allow the correct diagnosis to be made.

The timing of first feeding is thus of great importance and one rule of thumb for trout is that it should take place when between 10 and 25% of the fry swim up. With Atlantic salmon the problem is more complicated and although the experienced hatchery person may know intuitively when to start feeding, scientists generally advocate using the ratio of the weight of the yolk-sac to the embryo, i.e. first feeding when the embryo comprises 80% of the dry weight of the whole alevin.

A number of diseases affect fish when they first start feeding. In general it may be said that the younger the fish, the fewer symptoms are seen in any disease problem; the smallest fish simply die. Many of the causes of such loss are living agents, but consideration should first be given to two important non-infectious causes.

7.1 STARVATION

Even after fry have been successfully brought onto the feed, the hatchery person may be tempted to underfeed them, being aware of the dangers arising from uneaten food and detritus. However, this runs the risk of starving the fry, which may soon resort to cannibalism, those fish that have never really got onto feed often being the first victims. Large numbers of fish may die in this manner, and

Fig. 7.1 Starvation. Young fish that do not feed become progressively thinner until the head may appear to be disproportionately large, when they are referred to as pinheads.

individuals often lose their eyes in the initial attack. A fish that apparently has a tail at both ends is sometimes the first evidence of one fish attempting to swallow another. Starvation may be more easily avoided if care is taken to prevent overstocking and to ensure that the hatchery tanks are cleaned daily. Even if actual fish deaths are not occurring, poor feeding regimes and bad husbandry will increase the number of uneconomical pinheads. As a general maxim it may be said that for first feeders, the more the farmer tries to feed little and often, the better. Cannibalism is, of course, a major way in which any diseases affecting the poorer fry may be spread to the others.

7.2   GAS-BUBBLE DISEASE

Supersaturation of gases may cause losses among early feeders, as well as among alevins and older fish. The clinical signs are similar to those shown by alevins, i.e. swimming belly-up or vertically, and often affected fry will be seen to have a bubble of gas in their mouths. Wet preparations of gills will also show evidence of gas bubbles within the blood capillaries and such bubbles may be clearly visible in the fins and tail. Prevention is by aeration of the water supply and careful maintenance of pumps and pipes.

7.3   ACUTE LOSSES

Several infectious diseases of early feeders may cause sudden severe losses. These should not be confused with fish kills, since it is most unlikely that any infectious disease will result in more than 20% of the affected stock succumbing over a period of 24 hours, whereas fish kills frequently cause 100% mortalities within minutes. Twenty per cent losses occur typically with certain virus diseases and may be called an 'acute' loss. The same organism may, however, have far less dramatic effects under different circumstances, e.g. where fish are less susceptible because of age or acquired resistance, etc. Nevertheless, very young fish are particularly vulnerable, and heavy losses should cause the farmer to suspect one or more of the following conditions.

7.3.1   Infectious pancreatic necrosis (IPN)

In a fish farm which has never previously been affected by an outbreak of IPN, the cardinal sign of the disease, if one of the more pathogenic strains is involved, is the occurrence of high mortality in young fry during the first 2 months after coming onto the feed. Often losses start to occur about 6 weeks after first feeding. The affected fry move reluctantly and tend to swim on their sides or with slow spiral movements, frequently sinking to the bottom. They may be darker in colour and have swollen bellies (Plate 24). When killed and opened up, there is often whitish mucus in the gut but no food present. Occasionally there may also be small blood spots over the stomach area.

The virus responsible for IPN attacks the cells of the digestive system so that both the gut lining and the pancreas are destroyed. Definite diagnosis can only be made after cell culture and histology. If this disease is suspected, the farmer should place samples of the pyloric area of the stomach from freshly killed fry into 10% formal saline and send these to the appropriate laboratory, together with frozen specimens or live fish.

No treatment is possible and the farmer has to weigh up the importance of the affected stock to the farm's future programme. It is often advisable to kill out affected stocks (which may all die anyway). The entire fish farm may then be disinfected with an iodophor compound. Ideally live fish should not be moved to another farm, and eggs from fish on an affected farm should not be used for future hatchery purposes.

Adult fish may carry the virus within their organs, particularly the gonads, without showing any evidence of disease. The commonest mode of entry of IPN is in association with eggs brought in from an infected source. Wherever possible such eggs should be accompanied by a veterinary certificate that the farm has been tested for specific virus disease. This does not guarantee eggs as virus-free (the presence of very few virus particles, undetectable by tests, can still cause disease), but it greatly reduces the risk. In addition, it is advisable to disinfect eggs on arrival at the farm (see Chapter 13). IPN is a particularly insidious problem as it readily infects wild fish, and while these are not visibly affected, they act as carriers. Such carrier fish can reinfect the farm (e.g. by their excreta entering the inlet pipe) after it has been cleaned out and disinfected.

7.3.2 Infectious haematopoietic necrosis (IHN)

IHN is characterized by a sudden rise in mortality, particularly in young fish, under conditions of low water temperature, i.e. less than 10°C. Affected fish show pale gills, swollen bellies and may trail a ribbon of faeces from the vent (Plate 25). both salmon and trout are susceptible, and until recently the disease was confined to the USA, Canada and Japan, but it is now also present in mainland Europe. No treatment is possible and diagnosis is by sophisticated laboratory procedures as with IPN.

7.3.3 Costiasis

Two parasitic diseases can cause heavy losses of early feeders although neither are usually as sudden or severe as the viral diseases. *Costia* (*Ichthyobodo*) is a common cause of acute losses in young fish which usually show no other obvious symptoms. This parasite can cause a rapid build-up of infection in a very short time, particularly if water temperatures are rising, the fish are heavily stocked, and the tanks or ponds need cleaning. Diagnosis is by identification of the parasite in wet preparations of gills and/or skin. Treatment is by formalin, and earth ponds that suffer badly with *Costia* should be left fallow for a month in summer after liming.

Costia can be particularly important in first feeding salmon following spate conditions if hatchery waters have high particulate matter. The resulting hypersecretion of mucus allows rapid proliferation of *Costia*. Losses can be very high and treatment difficult.

7.3.4 Hexamitiasis (*Octomitus*)

Hexamita often causes chronic low-grade losses, but rarely if ever is it the cause of acute losses even in very small fish. Mortalities do not usually exceed 10% per day and they are associated with a reduction in appetite of the stock, trailing faecal casts and excessive nervousness. Sometimes affected fish show a red vent and a reddened gut when opened up. Diagnosis is by identification of the characteristic parasites in a wet preparation of either yellowish watery gut contents or bile from the gall bladder, usually in association with other diseases. Treatment is by furazolidone or alternative compounds added to the feed. This is often difficult since affected fish will not be feeding well and other underlying problems may still remain.

7.4 GILL PROBLEMS

Gill problems are not uncommon among early feeders, particularly where the standard of husbandry is poor. Characteristically such problems are manifested by several symptoms:

- continuous minor losses;
- fry gathering near the water inlets;
- fry 'riding high' on or near the surface of the water;
- obvious respiratory distress with gasping and puffed-out gill covers;
- a high incidence of pinheads contained in poor quality water.

These symptoms often follow grading or handling and may be more evident if a netfull of fry is placed in a bucket of water. The two main disease problems involved are bacterial gill disease and gill fungus, and both may exist together although one usually predominates. The precipitating conditions for these diseases centre around a poor environment such as is brought about by spate conditions, particularly in acid rain areas, or overfeeding, causing much suspended matter in the tanks or ponds. These solids tend to stick to the gill surfaces, and this suffocates the fish or irritates the gill surface and permits disease organisms to gain entrance. A variety of different organisms may be present, notably several species of bacteria (mainly myxobacteria), Protozoa (such as *Costia*) and fungi. In conjunction with the poor water quality (low oxygen levels and high ammonia levels), suspended solids and overcrowded conditions, these organisms bring about a disease complex which can arise overnight if fish have been stressed.

Examination of affected fish shows one of two different pictures. In the case of bacterial gill disease there is a mat of slimy bacteria coating the gills. When a wet preparation of gill material is examined under the 400× magnification, the swollen gill lamellae are obvious, and thread-like bacteria are present in large numbers, although individual bacteria may be difficult to see. Protozoan parasites are also seen frequently in association with the bacteria.

If gill fungus is predominant, the threads of fungus may often be visible to the naked eye. Under the microscope they are very obvious, surrounding and infiltrating the gill lamellae.

As well as trying to improve the environment of the fry, they may be treated by addition to the water of quaternary ammonium compounds (especially with bacterial gill disease), or malachite green (especially with gill fungus), provided these substances are available for legal use.

7.5 LOW-GRADE LOSSES

The daily occurrence of small losses among early feeders which are being adequately fed is usually due to a variety of mixed infections with external parasites. These parasites infest the skin and gills where they feed and cause irritation, which can sometimes cause symptoms usually associated with older fish, e.g. flashing (twisting over on their sides to give the appearance of a silvery shoal). If such fish are examined they often show an increase in mucus secretion, and severely affected individuals may show reduced appetite and the appearance of a rash on the sides.

Diagnosis is by identification of the particular parasites in skin scrapings and wet preparations of gills. The commonest parasites seen under these circumstances are *Costia*, the *Trichodina* complex, and the *Scyphidia* complex. Internal parasite infections, particularly with *Hexamita*, are also common under these conditions.

The external parasites may all be treated with formalin. More important, however, is correction of the husbandry faults which have allowed the condition to develop, namely improvement of water flow rates, decrease in stocking density, and the avoidance of undue stress, i.e. excess handling, dusty food and dirty fry tanks.

Chapter 8
Diseases of Growers in Freshwater

Among the commonest signs of disease is a reduction in appetite, coupled with the presence of fish that are dark in colour, lethargic, and swimming close to the side or outlet of the pond or tank. However, certain diseases may be identified by more specific symptoms. For example, aberrations of swimming behaviour may occur in several distinct forms. A disturbance of the swim bladder will cause mechanical interference with buoyancy. Similarly, nervous disorders can upset swimming patterns and cause loss of equilibrium, e.g. in virus diseases and in whirling disease. Disorientated swimming may be simply due to cataracts and consequent blindness. The presence of skin parasites may provoke irritation and result in curious swimming behaviour, e.g. if the fish attempts to scrape itself on the bottom of the pond, or exhibits flashing as it twists over in the water. These, and various other clinical signs, can often help the fish farmer to pinpoint the cause of a particular problem fairly quickly. For this reason a matrix has been constructed (Table 8.1) which lists the most significant diseases of growers in freshwater against their characteristic clinical signs.

These general features will apply equally to fish grown through their complete life cycle in freshwater or to Atlantic or Pacific salmon which will smoltify and transfer to saltwater. Since the smolting process is a particularly significant stage, diseases of smolting *per se* are considered in Chapter 9.

This matrix should not be regarded as a hard-and-fast set of rules, but may often be used to facilitate diagnosis. It will be seen that some signs, e.g. whirling, are generally an unequivocal means of identification, whereas others, e.g. inappetance or wasting, are far less specific. Reference should always be made to the detailed description of any condition under the appropriate section of this chapter before any form of diagnosis is made. The diseases of growers are listed under four main sections which are divided up according to the appearance of the disease on the farm. 'Gill problems' and 'skin problems' are self-explanatory; the distinction between 'acute disease' and 'chronic disease' is less obvious. In this text, acute diseases are those that are manifested by a sudden onset and usually spread rapidly throughout a particular stock. Chronic diseases are usually very gradual in onset and consequently often fail to be identified and remedied before the whole stock has become affected. In this case the pattern is similar to 'low-grade losses' of early feeders, although certain chronic diseases may be impossible to treat and can result in severe financial loss.

8.1 ACUTE DISEASES

8.1.1 Non-infectious diseases

As with other age groups of fish, if a problem arises extremely rapidly the cause is usually non-infectious in origin, as in a fish kill. Gas-bubble disease will also affect growers with varying degrees of severity, and the signs and remedies are the same as those given previously. A sudden reduction in oxygen levels which may follow a plankton bloom or release of silage liquors, or a sudden reduction in pH, or the presence of plant alkaloids from rotten leaves, following spate conditions, particularly in acid rain areas, may precipitate a major fish kill. However, frequently the fish may merely go off the feed and crowd the inlet or sides of the pond in an agitated fashion. The presence of small haemorrhages on the gills may be due to either reduced pH or to leaf toxins. The addition of lime (calcium carbonate) to the water will help to buffer any changes in pH.

The 'acuteness' of onset of an infectious disease problem among a population of fish is related both to the speed with which the causative organism can reproduce and to the fitness of the fish. Generally this means that viruses cause the most sudden appearance of disease symptoms, followed by bacteria, and then by parasites and fungi.

8.1.2 Virus diseases

Virus diseases can affect susceptible growers in the same way as early feeders, causing up to 20% losses within 24 hours. There are three main virus diseases likely to occur.

Viral haemorrhagic septicaemia (VHS)
Unlike the other virus diseases of salmonids, VHS usually only affects growers. The disease is a major problem in European trout farms, rainbow trout being very susceptible whereas brown trout appear to be more resistant. It has also been recorded in Pacific salmonids farmed in Europe and the USA and more recently in farmed marine fishes.

Affected fish generally exceed 5 cm in length, and the main sign is sudden, severe losses, usually associated with haemorrhages in the skin. Fish dying of VHS differ in their pathological features, depending on the stage of the outbreak. Fish that are completely susceptible appear very dark in colour and show rapid mortality. The gills are pale with red haemorrhagic spots and haemorrhages may also be seen around the eye. When opened (Plate 26), large blood clots are visible in the body fat, over the gonads and within the muscles. The liver is very pale while the kidney is thinner and bright red.

Table 8.1 Matrix showing the important diagnostic signs

	Cataract	Dark colour	Flashing	Gasping	Inappetance	Internal bleeding
IHN						++
IPN		+				+
VHS		++				+++
Bacterial septicaemias					+++	+++
Gill problems				+++	+	
BKD		+				+
PKD		+				
TB/*Nocardia*		+				
Costiasis			+++		+	
Eye fluke	+++	++				
Internal worms						
Large ectoparasites			+			
Hexamita (Octomitus)					++	
Trichodina complex			+			
Whirling disease		++				
White spot (Ich)			++			
Fungi (external)						
Fungi (internal)		+				
Fin rot		+				
Gas-bubble disease						
Fish kills						

After early mortalities in an outbreak showing the acute disease, the picture changes to the chronic form, where mortalities are lower and fish take longer to die. In this form the fish appear quite black and show severe popeye (Fig. 8.1). They are very anaemic due to the severe internal bleeding they have suffered and this is shown especially in the gills and liver; the swim bladder and kidney may be enlarged or the entire abdomen filled with fluid, producing a swollen, dropsical appearance (Plate 27).

Towards the end of an outbreak nervous signs appear and affected fish loop the loop with a slow tumbling movement. The presence of haemorrhages throughout the internal organs of salmonids should always make the fish farmer consider the possibility of VHS.

Outbreaks of this disease are most common when water temperatures are at their lowest. It is believed to be introduced by feeding infected marine fish in wet-fish diets. These should not be used. Although it has occasionally been reported among fingerlings in summer, the infection usually becomes dormant if the temperature of disease among growers in freshwater

Mortality	Motionless and nervous	Popeye	Skin lesions (excluding ulcers)	Swimming aberrations (excluding flashing)	Swollen belly	Skin ulcers	Wasting	Abnormal kidney
+++								
+++				+	+			
+++		++		++	+			+
++			+			++		++
++	+++							
+			+		++	+	+	+++
+					++			+++
+					+++		+	+
++								
							+	
					+		+	
			+			+		
+	+						+	
			+					
+				+++				
++			+					
			+++			+		
					+			+
			+			+		
++		++		++				
++++								

exceeds 8°C. In this case it will often recur when the temperature drops or the fish are otherwise stressed by handling, etc.

Since there is no treatment for any virus diseases, VHS must be prevented from entering a farm by prevention of wet-fish feeding, strict control of visitors, elimination of predators, and prior certification of egg and live fish introductions. If an outbreak occurs, it is usually best to kill out the entire farm, leave the ponds fallow and disinfect for 3 months before restocking with healthy fish. If more than one fish farm exists on the same watercourse, it is usually necessary to maintain a joint programme for prevention and control. VHS is subject to strict legislative controls in most countries with salmonid industries.

Infectious pancreatic necrosis (IPN)
Although IPN typically affects early feeders, it may also occur among growers, especially on a farm where the disease has become established. If sudden losses occur when growers are stressed at any time

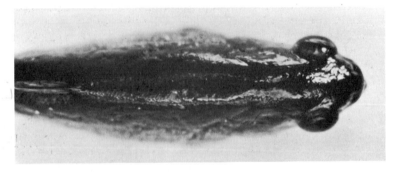

Fig. 8.1 Viral haemorrhagic septicaemia (VHS). This specimen is in the chronic stage of VHS, with darkened skin and marked popeyes due to haemorrhage behind them.

subsequent to an initial outbreak of IPN, then the possibility of a recurrence must be considered. The stress of transporting growers will often bring out the disease the following day ('transport IPN'), although grading, or even the onset of sexual maturity may be sufficient. Losses rarely total more than 20% of the stock and usually cease within 4 days, but surviving fish may grow poorly. When the guts of fish dying with transport IPN are opened, they are empty except for a coat of white mucus due to the effect of virus on the gut lining. In addition, haemorrhages may frequently be seen on the pancreas, stomach and associated intestine. In laboratory diagnosis of IPN infection, isolation of the virus from the gut, sex products or faeces of infected carriers (e.g. broodstock) is often made far easier if they have been transported or crowded immediately prior to testing.

Infectious haematopoietic necrosis (IHN)
The occurrence of IHN among growers follows the same pattern and should be dealt with in the same way as described for early feeders (see Chapter 7). It occurs at low water temperatures and is typified by sudden losses of fish which trail faeces from the vent and often show popeye, pale gills and swollen abdomens. When fish are opened, they show haemorrhages over the stomach area (Plate 28) which are more obvious than those that occur with early feeders (compare with IPN above).

Salmon pox
Although the virus thought to cause salmon pox has not yet been isolated, the condition appears very similar to warts, a virus disease of higher animals. In freshwater it is principally a problem of salmon parr during their first summer, and occurs as a series of often very disfiguring whitish growths on the body surface, particularly the flanks or back and the fins (Plate 29). The lesions do not appear to affect the fish significantly, and after a few months they slough off,

often suddenly, following handling of the fish. The ulcers that remain usually heal rapidly unless bacteria such as myxobacteria or *Aeromonas* become established (Fig. 8.2). If this happens severe mortalities can ensue, particularly if the water temperature falls, impairing the fishes' defences and wound-healing ability.

8.1.3 Bacterial diseases

Like viruses, bacteria may precipitate severe losses but the onset is not quite so sudden. The fish invariably go off the feed 1–2 days before they start to die. Bacterial diseases mostly occur at high water temperatures, and sudden inappetence under these circumstances is a strong indication that the farmer has to diagnose and treat the fish urgently, or severe losses will ensue.

Furunculosis
Furunculosis is caused by *Aeromonas salmonicida* and although characteristically a problem when water temperatures are at their peak, it can occur at most times of year. After a brief period of reduced feeding, fish will sometimes die due to furunculosis without showing any other signs at all. This is a particularly common pattern with small trout and Atlantic salmon parr, and can only be diagnosed with certainty by laboratory evidence of the bacterial colonies in the kidney and other organs.

In older fish, the disease can sometimes be identified by the frequent presence of furuncles. These are large, red, swollen, boil-like

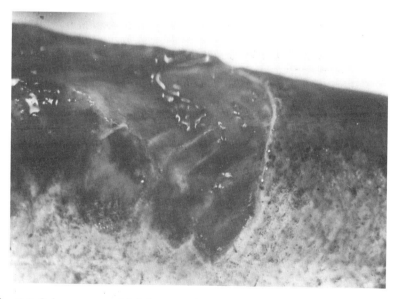

Fig. 8.2 Salmon pox which has eroded the epidermis as it sloughed and become infected with *Aeromonas hydrophila*.

lesions, which give the disease its name and are usually found over the shoulder or back (see frontispiece (d)). These may burst and release reddish fluid containing large numbers of bacteria which rapidly spread the infection (Plate 30). More often, however, among farmed fishes, the features of furunculosis are those of any bacterial septicaemia, i.e. darker colour, inappetence and occasional reddening of flanks or vent.

The bacteria involved persist in small numbers, between outbreaks of the disease, within the tissues of a few fish on the farm or in wild fish in the watershed. Such fish are known as carriers. The disease is more common in salmon and brown trout than in rainbow trout and particularly frequent in fish on farms that have wild migratory salmonids in their water supply. However, provided that affected fish can be persuaded to eat sufficient medicated food and that all severely affected fish, showing reluctance to feed or darkening or swellings, are removed and killed, all species can be treated by addition to the feed of various drugs. The advent of furunculosis vaccines and fallowing procedures have greatly reduced the incidence of furunculosis in farmed salmon (see Chapter 13 'Vaccination').

Enteric redmouth (ERM)

ERM gets its name from the reddish appearance of the mouth due to congestion of blood vessels, which occurs in a minority of cases, particularly in North America. However, the term 'redmouth' is something of a misnomer as the causative bacterium, *Yersinia ruckeri*, is also associated with general septicaemic diseases, especially when the water is enriched with high levels of organic matter. The disease is usually a problem with rainbow trout growers, which start to feed and grow well in early summer and then become listless, dark in colour and stop feeding. Losses of up to 20% may take place. ERM is frequently confused with furunculosis and affected fish usually show haemorrhages at the base of the fins and vent, with a congested or haemorrhagic appearance to the internal organs at post-mortem.

The disease probably occurs world-wide and can usually be treated with antibiotics given via the feed. However, where ERM is endemic and significant mortalities take place, or where fish are tranported live from one area to another, it is sensible to make use of one of the ERM vaccines which are now available.

Rickettsiosis

Rickettsiosis is a septicaemia of salmonids in Chile, originally only found in fish in seawater. Recently, however, it has also been detected in Atlantic salmon in freshwater. A similar *Ricke Hsia* has also been recently isolated from Atlantic salmon in Europe. Affected fish are dark in colour and go off their food, and have haemorrhages in the liver and a swollen spleen. There may be raised lesions in the skin also. Diagnosis is by demonstration of distinctive Gram-negative or Giemsa-positive cocco-bacilli in focal lesions in the liver and skin.

Other bacterial septicaemias

Apart from furunculosis, the *Aeromonas* and *Pseudomonas* bacteria are responsible for a variety of diseases in wild and cultured salmonids. In summer the diseases occur when water temperatures are high and there is a large amount of organic material present. Usually only a few fish are affected and the characteristic features are bright red blotches around the vent, back and sides. When opened up, such fish show haemorrhages throughout the internal organs and the kidney appears completely liquefied. A similar problem can occur later in the year and is called autumn aeromonad disease.

All of these infections may be treated by the use of antibiotics or sulphonamides in the feed provided that the fish will eat sufficiently, but improvement in water quality and aeration are important.

Botulism disease

Clostridium botulinum is notorious as an occasional cause of death in man and animals which eat rotten food containing its toxin. It was at one time also well known as an occasional cause of severe losses on some Danish trout farms using wet-fish diets during periods of high summer temperature. If the diet was not fed fresh under these circumstances, it may have begun to putrefy, allowing spores of *C. botulinum* present to produce their lethal toxin. The sudden massive mortalities resulting have caused the problem to become known in Denmark as bankruptcy disease Nowadays wet-fish diets are rarely used in salmonid culture.

However, botulism can also occur on trout farms unrelated to wet-fish diets. If putrefying trout carcasses are present on the mud bottom of ponds, they may be cannibalized by trout scavenging for food. *Clostridium botulinum* spores may well be present in the mud, and toxin formation can occur within the carcasses even at water temperatures below 10°C. Ingestion of the toxin by cannibal fish causes them to lose equilibrium and sink to the bottom. They will rest motionless on the bottom, then jerk themselves back to the surface again, but sink repeatedly and die within a few hours. Such fish will otherwise appear perfectly healthy on examination but must be immediately destroyed as they represent a very serious hazard to human health. Prevention is by prompt disposal of any dead fish and regular cleaning out of ponds. If botulism is suspected, the local veterinary and food health laboratories must be notified urgently and all fish sales stopped pending laboratory diagnosis of the problem.

8.1.4 Parasitic diseases

A number of parasitic diseases of growers can arise very quickly, particularly infections of the skin. Generally parasites cause reduction in appetite and signs of skin irritation or gill trouble. However, certain parasites can cause sudden losses without much prior warning. The parasites that may fall into this category are *Myxosoma*

(*Myxobolus*), *Ichthyophthirius* and PKD. Acute parasite infections with the first two organisms cause losses which may be associated with either frantic whirling movements (*Myxosoma*) or severe skin irritation, with white skin spots and attempts by the fish to scrape themselves on the bottom or sides of the tank or pond (*Ichthyophthirius*).

Whirling disease

The fast whirling movements which give this disease its name are caused by the spores of *Myxosoma cerebralis* and are quite characteristic. The spores of this parasite are formed in the cartilage of young salmonids, particularly rainbow trout, and cause a severe reaction when it changes to bone. In trout, cartilage becomes bone when the fish reach 6–8 cm in length and heavy infections will occasionally result in severe mortalities at this stage. More commonly, fish survive but show chronic damage. The spores damage the balance organ which causes whirling, and also press on the spinal column, which causes deformities of the back and tail and dark coloration (Fig. 8.3).

If whirling disease is suspected on the basis of these clinical signs, microscopic identification of the spores may be attempted. After decapitation of affected fish, the hardening cartilages of the skull should be scraped onto a slide. A wet preparation may then be made and examined under 40× magnification. Presence of the small shiny spores (see Plate 7) will give a positive diagnosis, but several fish may require to be examined. In many countries whirling disease is notifiable, hence if it is suspected an official laboratory should always be informed.

Fig. 8.3 Whirling disease. These fish have all suffered skeletal deformity as a result of the activity of *Myxosoma cerebralis* parasites. (*Courtesy of Professor N.O. Christensen.*)

This disease is prevented by rearing fish in water (e.g. spring water) that is free of the infective stage, which lives within nematode worms in the mud of the pond. If this cannot be done, then the use of concrete or fibreglass rearing tanks instead of earth ponds will minimize the chances of infection. Once their bones have hardened, salmonids are resistant to the signs of infection and can then be placed in earth ponds without risk of clinical disease.

White spot (Ich)

Acute infections with *Ichthyophthirius* can occasionally cause sudden losses in salmon and trout under conditions of high water temperature and low flow rates. Affected fish show extreme skin irritation and may jump all over the pond as if short of oxygen. They often scrape themselves along the bottom and sides of the pond and may be seen flashing. Examination of such fish usually shows the presence of small white parasites on the skin (Fig. 8.4). Skin scrapings of affected fish should show both the small rapidly revolving infective stage and the large brown adult with its horseshoe-shaped nucleus (see frontispiece (a)).

Ich parasites thrive in conditions of dirty water, low flow rates and dusty feed. The disease is particularly prevalent when water temperatures are above 17°C. Prevention should aim at good husbandry conditions and treatment is either by formalin or more preferably by a mixture of formalin and malachite green, if malachite green is legally available for use.

Proliferative kidney disease (PKD)

PKD has become recognized as one of the most significant of all diseases of salmonid culture. PKD principally affects rainbow trout

Fig. 8.4 White spot or Ich. *Ichthyophthirius* infection produces small, raised, whitish spots such as can be seen on this rainbow trout. Smears taken from such lesions show the characteristic parasites seen in Fig. 4.3.

but also occurs in brown trout and occasionally salmon. Although it is poorly understood, the parasite, in some still unrecognized form, appears to invade the fish in spring and early summer, particularly at the fingerling stage. It only appears in fish on riverwater and it is possible that another host, in the river, also plays a role in the life cycle.

Affected fish have a greyish swollen kidney and spleen and usually have pale gills due to anaemia. They may also have swollen and protruding eyes and a swollen belly containing clear fluid (Fig. 8.5). Such fish do not feed well and appear listless. The course of the infection is approximately 3 months and during this time losses may vary from a very small number to almost 100% depending on the severity of the infection and, particularly, on the husbandry. Fish with PKD must not be disturbed in any way and feed levels must be low. Any attempt to grade or move them or even to feed antibiotics can result in catastrophic losses.

Some farmers believe that exposing fingerlings briefly to infected riverwater early in the season will confer a degree of subsequent protection, especially if they are also treated with malachite green. However, given current safety concerns over the use of malachite green, there is no reliable means of control or treatment, although conservative stock management during the critical period of PKD infection can help to keep losses down.

Hexamitiasis (*Octomitus*)
Unlike the other parasitic diseases, infection with *Hexamita* often

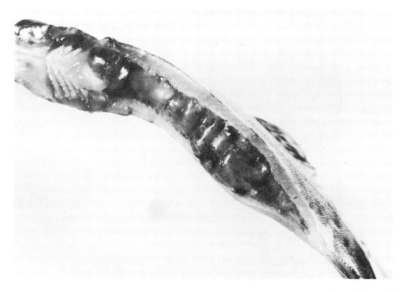

Fig. 8.5 The grey shiny lesions in the swollen kidney of a fish affected by PKD require careful differentiation from BKD. (*Courtesy of Dr H. Ferguson.*)

does not show distinctive signs other than possibly a sudden increase in losses. Indeed, many authorities consider this parasite to be merely a commensal in the intestine of fish that are not thriving for other reasons. The parasite is in the intestine where it causes haemorrhage which may sometimes result in an obviously reddened gut and vent. Mortalities do not usually exceed 10% per day and the stock may show excessive nervousness and go off their food. Less affected fish are dull, often dark-coloured and have a reduced appetite.

When the intestine of such fish is opened it may contain a clear or yellowish fluid (compare this with IPN). The parasites may be seen in wet preparations of the fluid (see Fig. 4.4). Antibiotic treatment may reduce levels of infection, but usually no treatment is necessary.

8.2   SKIN PROBLEMS

The external surface of fish includes parts of the gills and the eye and is often a very good indication of the state of health. Fish that are unwell are less able to resist invasion of the external surfaces by various parasites and fungi. Some viral and bacterial infections are also manifested by changes in the skin of the host.

Once again one may broadly divide the diseases affecting the skin into two main groups: those that result in a sudden acute problem and those that cause a chronic problem.

8.2.1   Acute skin problems

The vast majority of acute skin problems in salmonids are due to external parasite infections. These irritate the fish which may respond in a variety of different ways.

Flashing
Flashing is a very common phenomenon among growers and describes the situation when one or more of a group of fish suddenly make a rapid movement on their sides, momentarily giving a silvery appearance under the water. This sign, when repeated, should encourage the farmer to examine several netfulls of fish. If these show evidence of an increase in mucus secretion (often blue-grey in colour) and the presence of irregular pale patches on the skin, then the cause is likely to be an external parasitic infection. Often, however, such fish may flash without other signs, apart from a slight reduction in appetite. In these cases it is necessary to make a wet preparation of several skin scrapings. Microscopic examination of 40× magnification will permit identification of the commonest cause – *Costia*. Lower power observation (10×) will permit differentiation between the other main causes of flashing. In order of increasing size, these comprise the organisms of the *Trichodina* complex, of the *Scyphidia* com-

plex, *Ichthyophthirius* and *Gyrodactylus*. The last two usually cause other signs and are described in the next section. *Scyphidia* causes only mild irritation, but heavy *Trichodina* infections may cause a rash on the skin (Fig. 8.6). Affected fish in this condition may be seen swimming up to the surface and then dropping down to the bottom, apparently lifeless (similar to fish with botulinum disease, but resulting in much smaller losses).

External parasitic infections are most commonly seen in ponds with a low water exchange rate. These parasites thrive in conditions of poor water quality, especially if the fish are overstocked and there is much suspended matter present in the water, e.g. due to dusty feed. Thus control measures should include rigorous attempts to improve the husbandry conditions, e.g. by thinning out the stock in the ponds. However, some skin parasites, notably *Costia*, will sometimes cause trouble even when conditions are excellent. In these cases it is necessary to use a formalin treatment, given as a flush or more preferably a bath (see Chapter 13, 'Treatment methods').

Scraping

In certain types of severe skin irritation, affected fish may be seen to scrape themselves along the bottom or bank of the pond. Such fish frequently show other signs such as flashing, jumping at the inflow, reduced appetite and varying degrees of mortality. This combination of signs is invariably due to one or more of the following parasites: *Ichthyophthirius*, *Gyrodactylus*, *Lernaea* and *Argulus*.

All four infections are fairly easily identified without the need for microscopy. *Ichthyophthirius* is a particular problem under conditions of high water temperature and usually causes pustules on the skin. *Gyrodactylus* may be recognized as its worm-like body changes shape

Fig. 8.6 External parasitism. *Trichodina* have caused this lesion on the back of an Atlantic salmon. The whitish patch (arrowed) is a mixture of parasites, tenacious mucus and skin cells.

on the fish or after transfer to a microscope slide. *Lernaea* usually inserts its anchor at the base of a fin or in the vent area, while *Argulus* moves rapidly around the fish and will often swim away as soon as affected fish are netted out for examination.

As with other external parasites, the first condition of effective control is to eliminate the environmental factors that predispose to the problem. However, all four infections may be treated by the addition to the pond of various chemicals. *Ichthyophthirius* is best treated by a mixture of formalin and malachite green or, as with *Gyrodactylus*, by formalin alone; *Lernaea* and *Argulus* by careful treatment with one of the newer more environmentally friendly organophosphorous compounds. In addition, it is advisable to leave badly affected ponds fallow for a month after ploughing in quicklime. This is best done in summer and may also be of use when *Costia* is a problem.

Saddleback

Bacterial infection of salmonid skin with *Flexibacter columnaris* results in the typical saddleback lesion of columnaris disease. This is usually a disease of high temperatures and affected fish appear dull, inappetent and have a greyish sheen along the back. This usually incorporates the dorsal fin which gradually becomes eroded.

The grey mat of bacteria, mucus and skin cells also begins to erode the epidermis and then to cause sloughing of the skin until the muscle is exposed. The lesion then usually has a bright red, inflamed centre with a grey rim. This extends across the back to form the 'saddle' which gives the disease its name. Diagnosis can be confirmed by examining a wet smear of the edge of the lesion, when the very obvious columns of *Flexibacter* bacteria can be seen.

Although antibiotic treatment may reduce losses provided the fish are feeding, it is the stress of the very high temperature that allows the infection to take hold. Falling water temperatures usually resolve the problem quickly if the fish survive.

Sunburn

Sunburn is a serious disease in both trout and salmon. It occurs in the wild but is principally a problem with farmed fish, and it occurs in two forms. The first is straightforward exposure to the traumatizing effects of ultraviolet light in fish that are held in clear water for any length of time. Salmonids do not normally have any protective pigment within their epidermis so they are particularly vulnerable. Whereas wild fish in natural conditions would move into deeper water or into shade when sunburn is likely, farmed fish cannot do so because they are constrained by a cage or tank.

The condition is particularly important in fish farmed at high altitude, in the southern hemisphere where ozone layer thinning occurs, and in fish supplied with borehole or well water which have very little in the way of suspended solids to provide a filtering effect.

Affected fish may have whitish lesions on the top of the head, fins or tail (Fig. 8.7). They usually develop fin rot, and bacterial infections readily overlay the original damage. The only treatment is to provide shading for the affected fish, especially during grading.

A less common but potentially more dangerous form of sunburn is the condition known as photosensitization. This occurs when the diet contains one of the 'photosensitizing' chemicals. When these compounds are stimulated by even low levels of ultraviolet light they release energy, and if this happens in the blood vessels of the skin the skin 'burns' and peels off. This is probably more common than is realized, and several serious epizootics have been reported.

Usually the photosensitizer is a chemical that has been accidentally incorporated into the diet during manufacture, but certain treatment drugs can also be toxic in this way, as can some natural products. The

(a)

(b)

Fig. 8.7 Sunburn lesions. (a) Bald spot lesion in American lake trout (*Salvelinus namaycush*). (*Courtesy of Dr L.N. Allison.*) (b) Generalized sunburn with secondary infection in young salmon parr. (*Courtesy of Dr A.M. Bullock.*)

level of ultraviolet light required to stimulate photosensitization is low, so shading is unlikely to prove more than a mild palliative. The only means of control is to prevent incorporation of the photosensitizing agent in the first place.

Others

A variety of other acute conditions can affect the skin. Predators, such as herons, frequently damage fish, causing wounds and ulcers which may become secondarily infected with bacteria and fungi. Ulcers and haemorrhages may be due to acute bacterial infections such as coldwater disease or furunculosis. Virus diseases (e.g. VHS) can cause skin haemorrhages, but one such disease, salmon pox, causes proliferation of warts over the skin surface. A skin disease of rainbow trout confined to certain Pacific areas of the USA is 'strawberry disease'. Its name comes from the characteristic red, lumpy, ulcerated lesions occurring on the skin of affected fish. The condition usually commences in autumn, reaching a peak in late winter, when up to 80% of fish may develop it. Mortalities are minimal but affected fish cannot be sold. Although the cause is not known, treatment with antibiotics reduces both the seriousness of the disease and the usual 8-week course of an infection.

8.2.2 Chronic skin problems

Many of the acute skin problems if permitted to continue will cause the fish to become increasingly emaciated and undernourished. Such fish are usually dark in colour and lethargic. The skin of these fish initially secretes excess mucus. Later, however, it becomes rough, discoloured and readily invaded by other organisms, especially fungi.

A frequent occurrence, particularly under conditions of poor husbandry, is that of fish with eroded tails and fins. At first the fins merely show a ragged edge (Fig. 8.8), but gradually they become progressively more eroded. At this stage bacterial and fungal colonies have usually become established and they extend the process into healthy tissue which subsequently dies. Eventually the infection becomes generalized resulting in death. It is no use simply killing off these organisms in order to reduce the incidence of fin rot and peduncle disease (Plate 31). The disease process is often primarily due to the response of the fish to an insanitary environment, sometimes complicated by an unbalanced diet. Adequate water flows and clean ponds or tanks, which are not overstocked, will prevent these problems becoming significant.

One particular problem with salmon parr populations is dorsal fin rot. This is initiated in early autumn when reduction in water temperature leads to the majority of the stock moving to the base of the water column. Here territorial aggression leads to damage and secondary infection of the dorsal fin. Apart from its unsightliness, dorsal

Fig. 8.8 Fin rot. This myxobacteria lesion depends on some predisposing factor such as overcrowding, dietary imbalance or poor water quality.

fin rot can also act as an important portal of entry for bacteria such as *Aeromonas salmonicida*.

8.3   GILL PROBLEMS

Gill problems are not nearly so much of a problem in growers as among very young fish. However, poor husbandry, which is usually manifested by skin problems, will also increase the incidence of gill diseases. Skin parasites and related organisms will often colonize the gills and such fish may show obvious signs of respiratory distress. In these circumstances, affected fish tend to crowd the inlets and show extended gill covers and rapid gasping. Such fish usually have a reduced appetite and suffer variable losses, particularly when water temperatures are high. For accurate diagnosis a wet mount of the gills of affected fish should be examined. The commonest causes of such problems are small ectoparasites (e.g. *Costia*), gill flukes (notably *Gyrodactylus*), and various myxobacteria. However, apart from the specific instance of the primary pathogen *Gyrodactylus salaris*, which is notifiable, the primary cause is the poor environment.

Therefore the use of formalin (against *Costia* and gill flukes) or quaternary ammonium compounds (against myxobacteria) without appropriate correction of the husbandry conditions is at best a short-term measure and at worst the extra stress which finally kills the whole stock in a pond.

A different sort of problem is posed by the parasitic larvae of the freshwater mussel which, if released into the water near to a hatchery, can result in severe gill damage, especially to salmon parr. The glochidia may be released in very large numbers and can cause spectacular losses from the damage which they inflict on the gills of the affected fish. Treatment is impossible and since freshwater mussels are, in most countries, rare and endangered animals, and are often legally protected species, they must not be disturbed. If an alternative water supply is not available for the short period of maximal glochidia release then a fine gravel filter or recirculation system will have to be installed to reduce levels of gill invasion.

If gill problems are significant under conditions of satisfactory husbandry, a careful check should be made on the water chemistry. The most important cause of gill damage associated with changes in water chemistry is the acid rain effect, usually occurring in soft water and particularly associated with snow melt and rainfall run-off from coniferous forests (Chapter 11). But the presence of certain effluents, notably suspended solids and dissolved heavy metals, will provoke severe gill problems and abundant mucus production. This situation may not cause concern when water temperature is low and flow rates are high. However, when the temperature increases, with consequent reduction in dissolved oxygen levels, such hidden problems may suddenly worsen the already inadequate gill function, causing severe stress and even heavy losses.

8.4   CHRONIC DISEASES

If a population of fish survives an acute outbreak of disease, frequently a proportion will become chronically affected. Thus fish that have survived an initial outbreak of whirling disease will continue to exhibit the signs of infection. Carrier fish which have survived an outbreak of IPN virus are usually vulnerable to the stress of grading or transport. With VHS disease, the appearance of nervous signs indicates that the acute phase of the infection has given way to a chronic phase. In this case the chronic disease state heralds the end of the disease outbreak. In the case of certain diseases, however, the infection is essentially a chronic process from the start. Human TB and leprosy are examples of such chronic infections and similar infections occur in fish. In this section, chronic diseases are divided into these specific diseases and into a group called 'bad-doers'. The latter embraces the various types of chronically diseased fish which are a common occurrence on many fish farms. Their poor state of health is usually evident over a considerable time and can have a variety of causes.

8.4.1   Chronic infections

Certain chronic infections such as whirling disease and fin rot have been discussed earlier. Of the remainder, five specific infectious diseases are important. Three of these are parasitic infections (eye fluke, *Hexamita* and *Diphyllobothrium*) and two are caused by bacteria (BKD and TB). Eye fluke and *Hexamita* infection are quite common diseases and BKD can in certain circumstances be one of the most economically important of all the salmonid diseases.

Bacterial kidney disease (BKD)

In freshwater BKD is characteristically a disease of late winter and spring. Fish that have survived winter in reasonable condition suddenly show reduction in growth and conversion efficiency, with spasmodic, then increasing, mortalities. For some unexplained reason it appears particularly prevalent in cage culture, but can also occur in pond, tank or raceway.

Affected fish are usually dark with swollen abdomens, and may have popeyes and small haemorrhages around the pectoral fins and the vent. If they are cut open there may be white nodules dotted around the spleen, liver and kidney, especially in Atlantic salmon. However, large grey lesions may also be seen in the kidney, spleen and through the muscle, where cavernous spaces may occur, particularly in Pacific salmon (Plate 32(a)). In very chronic cases the fish may look silvery in external appearance, whereas internally a grey membrane often forms over the abdominal organs, and red splashes of haemorrhage may be seen on the body wall. Often only a small number of fish are affected, others appearing perfectly normal.

The disease is caused by a very small bacterium, *Renibacterium salmoninarum*, which is thought to be transmitted via wild fish, especially Atlantic salmon, in which natural epizootics are not uncommon (Plate 32(b)). A fish farm on such an infected watershed is therefore very vulnerable to infection, but an equally significant means of infection, even where fish are reared on borehole water, is vertical transmission via eggs from infected carrier broodfish, or infected fry which are brought onto the farm. Infected eggs can carry the bacterium within the yolk, so that routine disinfection procedures will have no effect on its transmission.

Equally, no treatment has been shown to have any curative effect on fish. It may be possible to mitigate the scale of losses by conservative husbandry coupled with high levels of trace elements in the feed. Where the infection is restricted to one particular stock origin of fish in a cage population, it is worth killing out the affected cages as the remainder will often remain free of disease.

Plate 27 Viral haemorrhagic septicaemia (VHS). This is an example of the chronic stage of VHS, showing the popeyes and swollen, grey-coloured, nodular kidney. (Courtesy of Professor N. O. Christensen.)

Plate 28 Infectious haematopoietic necrosis (IHN). Although predominantly a disease of very young salmonids and affecting the kidney and spleen, the virus of IHN also affects the pancreatic tissue as can be seen in the pinpoint haemorrhages on this specimen (compare this with IPN). (Courtesy of Dr T. Kimura.)

Plate 29 Salmon pox forming a saddle over the back of a one-year-old Atlantic salmon parr. (Courtesy of Dr T. Håstein.)

Plate 30 Furuncle dissected to show the necrotic muscle which contains very large numbers of *Aeromonas salmonicida* bacteria.

Plate 31 Peduncle disease. This case was associated with precocious sexual maturity in fish held in saltwater. The tail has become infected with myxobacteria, but in freshwater the same condition often predisposes also to *Saprolegnia*.

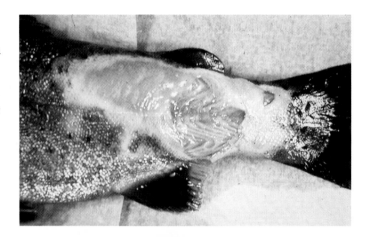

Plate 32 (a) Severe BKD. In Pacific salmon culture especially there is frequently muscle involvement in BKD. The cavernous lesions of the muscle are in addition to the small white granulomas in the kidney. (Courtesy of Professor R. D. Wolke.)

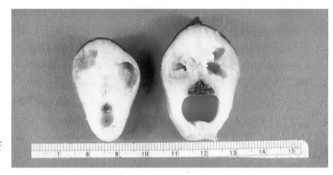

Plate 32 (b) The whitish lesions of BKD are seen in the kidney substance of this infected Atlantic salmon smolt. (Courtesy of G. Macgregor.)

Plate 33 Brown trout with heavy *Diphyllobothrium* plerocercoid infection of the viscera. (Courtesy of Dr R. Wootten.)

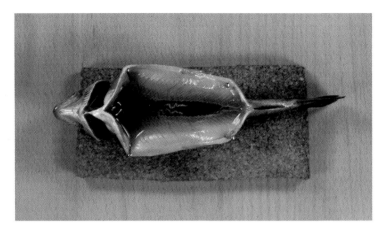

Plate 34 Nephrocalcinosis in rainbow trout. The ureters on the surface of the kidney are blocked with whitish granular deposits.

Plate 35 Dissection of the enlarged swim bladder of a trout with bloat associated with blockage of the pneumatic duct with dusty feed. (Courtesy of Dr S. Mohammed.)

Plate 36 Mass mortality of market size salmon following a red tide bloom.

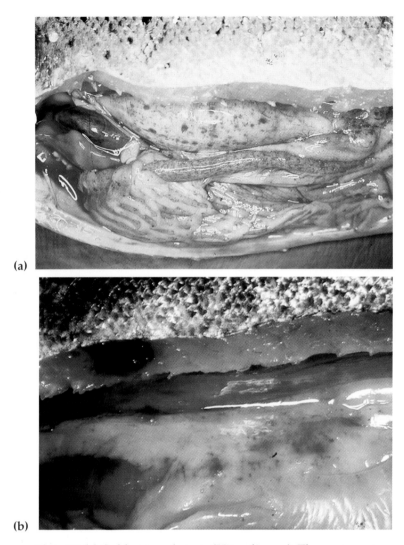

(a)

(b)

Plate 37 (a) Coldwater vibriosis (Hitra disease). The post-mortem picture in *Vibrio salmonicida* infection is characterized by haemorrhages of muscles and viscera. (b) *Vibrio anguillarum* affected fish showing the deep-seated muscle lesions. (Courtesy of Dr T. Håstein and Dr B. Hjeltnes.)

Plate 38 Coho salmon with typical skin lesions associated with rickettsiosis *Piscirickettsia salmonis* infection. (Courtesy of E. Branson.)

Eye fluke

Fish with eye fluke (*Diplostomum*) infection show a whitish opacity or white specks in one or both eyes, and varying degrees of blindness (Fig. 8.9). They are usually dark-coloured and often swim at the sides of the pond, where they are easy prey for predatory birds.

The infecting stages of the eye fluke, the cercaria, are released from snails in largest numbers in May and June. Where infection burdens are heavy, the invading cercaria cause severe irritation along the lateral line area on each side of the fish, as they penetrate in their hundreds. This can lead to extreme irritation, ulceration, secondary bacterial infection and death.

The parasitic life cycle involves snails and fish-eating birds. It may be broken by the use of molluscicides against the snails. Electric grids may be placed across the water inlet in an attempt to kill off the infectious stages after they have left the snail. The use of predator netting, bird scarers, etc. will discourage birds from either eating infected fish or transmitting infection to the farm in their faeces. Diagnosis can be confirmed by microscopic examination of a smear of the macerated lens (Fig. 8.10).

Nutritional cataract

Occasionally freshwater diets, especially for salmon parr, may have excessive levels of ash in them or be actually deficient in zinc. They also occasionally have vitamin A or vitamin E levels that preclude proper metabolism, and in both cases cataracts may develop in the eyes. These are generally an indicator of the deficiency rather than a

Fig. 8.9 Eye fluke. A cataract, or opacity of the lens, can be seen through the pupil and is due to parasitic infection with *Diplostomum*.

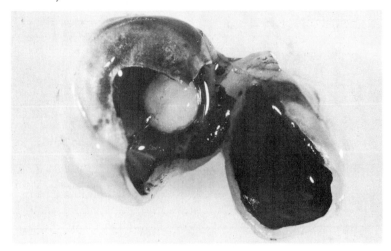

Fig. 8.10 Dissection of the eye of a rainbow trout with eye fluke. The lens is obvious as a white ball bathed in fluid. When it is chopped up and made into a smear preparation the parasites may be seen (see Fig. 4.17).

serious problem *per se*, since unless they are very severe, the fish can still see sufficiently well to feed.

Hexamitiasis (*Octomitus*)

Hexamita causes an internal parasitic infection of salmonids, which often show few distinctive signs other than an increase in losses, reduced appetite and excessive nervousness. The parasite is in the intestine, and when affected fish are opened up this usually contains a clear or yellowish fluid. Diagnosis is by identifying the parasites microscopically using a wet preparation of gut contents.

Treatment can be achieved by using furazolidone or other compounds in the feed. However, the occurrence of *Hexamita* infection in growers is usually indicative of some other underlying cause of ill-health allowing the parasite to become established as a secondary infection. Thus, treatment for *Hexamita* should be accompanied by a comprehensive disease investigation to determine what other problems may be present.

Diphyllobothrium infection

Diphyllobothrium infections of trout and young salmon grown in freshwater cages have become so significant that some freshwater lakes cannot be used for cage culture. The infective stage of the parasite is consumed, within its copepod host, by rapidly growing fish and the resulting parasitic cysts can cause destruction of the abdominal viscera. Affected fish are poor growers, and where heavy infections exist, serious marketing problems ensue because of the appearance of highly active tapeworms at evisceration. There is no suitable treatment for the condition and it is usually impossible to control the aquatic birds which act as the final host. It may therefore be necessary to avoid stocking freshwater cages when birds are in greatest abundance (e.g. the nesting season) or to moor the cages beside the water inflow to the lake (Plate 33).

Young salmon parr reared in freshwater cages can have their growth stunted by heavy *Diphyllobothrium* infection, but when transferred to seawater they generally grow satisfactorily and resolve the chronic lesions over time.

Tuberculosis (TB) and *Nocardia* infection

Infections with either *Mycobacterium fortuitum* or *Nocardia* bacteria are the cause of swellings in the head and abdominal regions together with generalized wasting. Both infections can affect whole stocks of fish which will initially show a reduction in conversion efficiency and then gradual development of lumps on the body and increased losses (Fig. 8.11). If affected fish are cut open, they will usually show either one large white lump or many small white specks scattered throughout the body (Fig. 8.12).

Tuberculosis is commonly associated with the feeding of trash fish or fish offals. Severe outbreaks have occurred when viscera from salmon and trout have been fed back to the fish without prior pasteurization. Treatment is usually not feasible and control is by changing the method of food preparation.

8.4.2 'Bad-doers'

The occurrence of fish that 'do badly' may usually be directly related to the husbandry conditions under which they are kept. Fish show-

Fig. 8.11 *Nocardia* infection in Pacific salmon. The swellings produced by this bacterial disease, known as granulomas, are found in the mouth or abdomen and produce severe deformity. (*Courtesy of Professor R.D. Wolke.*)

Fig. 8.12 *Nocardia.* At post-mortem the whitish structures containing the bacteria are seen. The lesions are very similar to those of *Streptomyces* and TB infection. (*Courtesy of Dr S.F. Snieszko.*)

ing fin rot, and various external parasite infections, lose condition and frequently become secondarily invaded by bacteria and fungi (Fig. 8.13). When such fish are cut open, they often show signs of internal parasitism. Roundworms, tapeworms and thorny-headed worms may be present in large numbers, and protozoan parasites, particularly *Hexamita*, may also be evident in wet preparations of gut contents. These parasites multiply rapidly when the resistance of the host is poor and bad-doers will often provide a focus of infection for various agents. The infection may then spread to healthy fish, and although these may not show obvious wasting or permit the parasites to multiply extensively, the conversion efficiency of the stock may fall significantly. Another consideration is that the presence of worm parasites can substantially reduce the market value and aesthetic appeal of fish, particularly if destined for human consumption.

Poor nutrition is of course also a possible source of poor doers, and as well as frank deficiency or bad feeding regimes, poor storage of feed can readily induce marginal deficiencies which are difficult to detect. Pansteatitis (see Chapter 10) is one particularly significant nutritional disease of growers which can result in very high mortalities.

Three specific chronic, non-infectious conditions related to husbandry which cause poor performance on certain trout farms are visceral granuloma, nephrocalcinosis and bloat. For the purposes of this section, visceral granuloma and nephrocalcinosis will be considered as a single entity.

8.4.3 Visceral granuloma/nephrocalcinosis

Visceral granuloma and nephrocalcinosis are probably different forms of the same disease condition of trout. They are characterized by losses occurring over a considerable period and any stress may increase the scale of losses. Visceral granuloma, which is particularly prevalent in brook trout, affects the stomach wall, while nephrocalcinosis affects the kidney. Both seem to be associated with

Fig. 8.13 A bad-doer. Where stocking densities are high or water quality poor, ragged fins can be a sign that the fish are under stress, and bacterial or parasitic problems may soon follow.

poor water quality (in particular high carbon dioxide levels) and possibly also dietary factors.

Changes to the stomach involve small hard raised lumps on the stomach wall which extend to other abdominal organs. In the kidney the lesions occur as long grey strands of hardened tissue extending into the kidney from its surface (Plate 34). Care should be taken to distinguish the condition from BKD.

Bloat

Bloat or over-inflation of the swim bladder is not uncommon in rainbow trout but much less frequent in other species. The condition can reach high levels, particularly when fingerlings are first brought outside from the hatchery. Of the two forms known, the more common form results from obstruction of the pneumatic duct, the small canal connecting the swim bladder to the back of the mouth (Plate 35). This can be caused by parasites or, more frequently, by particles of dusty food. Gas cannot leave the swim bladder and slowly accumulates to increase the buoyancy of the fish, causing it to float on the surface in a head-up or tail-up position and swim frantically in an effort to sink. Eventually such fish will die, but some can be treated by inserting a hypodermic needle into the swollen flank and releasing the gas. A proportion will simply re-inflate, but many can recover.

The other type of bloat is associated with overfeeding rainbow trout in winter. The feed accumulates in the intestine, but is poorly digested. The mass of food displaces the swim bladder which may develop a twist, preventing air exclusion from the posterior part. The latter then swells and the fish generally float tail-up. This condition can affect a very large number of fish in a pond, but usually a high level of cure can be achieved by ceasing feeding for several weeks, then commencing to feed again at a very low level.

Chapter 9
Diseases of Growers in Seawater

Marine culture of salmonids has developed very rapidly during the last decade and scientific advances have greatly helped in the control of infections. The disease problems of marine culture closely resemble those seen in freshwater and many of the infectious agents involved (e.g. *Aeromonas salmonicida, Costia*) can be pathogenic to fish in both types of environment. Other organisms, such as *Vibrio anguillarum*, can normally only survive and cause disease under marine conditions because of their particular need for salt. Nonetheless, the general principles of fish husbandry and disease control are common to both freshwater and marine conditions. Thus, sudden mass fish kills in a marine farm can take place for all the same reasons as in freshwater, e.g. algal blooms and oxygen lack, although the algal species are different and the oxygen solubility of seawater is less than freshwater. Because of the latter, oxygen lack can be a particular problem with marine cages at periods of slack water or if the effective stocking density is suddenly increased when tidal movements deform cages. Moreover, salmonids at sea need to excrete high levels of salt taken in when drinking and this physiological problem has important implications for the marine farmer. Thus, any further salt intake causing rapid dehydration and the loss of scales, from careless handling or fish becoming bagged in net pens, can result in large and sudden losses. The most critical period for salmonids in marine culture is during acclimatization to saltwater and this is now discussed in more detail.

9.1 SEA TRANSFER PROBLEMS

During the period of sea transfer the fish are under stress, being required to make major adjustments to control salt levels in the blood. Atlantic and Pacific salmon go through a complex series of hormonal changes around this period and unless they are transferred from freshwater to seawater at the right time, they may be physiologically unable to survive and adapt to the new environment. This osmoregulatory failure is particularly likely to occur if they have also been stressed by the effect of recent chemical treatment on the gills, by sudden changes in water temperature at the hatchery, or by the effect of long distance transport.

If salmon smolts go into osmoregulatory failure on transfer to seawater, the usual picture is of obvious stress resulting in very high losses during the first 2–3 days after arrival. Unlike normal smolts, the fish do not shoal properly but move listlessly without feeding and many appear very pale in colour. On closer inspection, signs of muscular tetany may be evident as a rippling along the flanks, and increased permeability of the cornea often results in severe cataract and blindness due to lens oedema. This eye effect is particularly likely if fish have been sedated during transfer. Depending on circumstances a certain proportion of the fish may settle down and osmoregulate satisfactorily whereas if the affected pens can be moved into less saline water, even some of the obviously stressed fish can recover, with reversal of the cataract and other signs. The key to avoiding such osmoregulatory problems is to carry out preliminary small-scale transfers of smolts as a trial routine and, if in doubt, to test the salt tolerance and the survival of individual smolts in laboratory experiments before transport begins.

Up to 2% of smolts normally die during the first week at sea even when the timing of transfer is correct. Many of these mortalities are due to fungal infection occurring at the hatchery just prior to transport. The hormonal changes of smoltification predispose smolts to fungal infection because of resulting skin changes and its incidence markedly increases towards the end of the transfer season. Even a very small tuft of *Saprolegnia* fungus, located typically at the base of the pectoral fins, can result in waterlogging and death. Of the remaining 'normal' transfer mortalities, there will often be a number of salmon parr that have escaped detection at the hatchery, together with precocious males and fish damaged in transit. If the cage is towed back to its moorings too quickly, or caught in cross-currents just after receiving the new batch of smolts, bagging problems can easily occur with fish dying from consequent scale loss.

A very poorly understood term, the 'fading smolt syndrome', is sometimes used to explain greater than normal levels of mortality after transfer of smolts to seawater. There is unlikely to be one single cause, and often problems associated with inadequate levels of feeding of the fast growing fish, possibly linked to other husbandry problems, are to blame. At other times, acute transport IPN or transfer of smolts that have not been fully acclimatized to saltwater existence may be the explanation of these losses.

Newly transferred fish are particularly at risk from infectious agents, and even sea lice attacks can cause large losses before treatment is possible if the resulting irritation causes excessive jumping, abrasion and scale loss. Any diseases that the fish may have been incubating or carrying without any effect in freshwater can quickly become clinically apparent. Fish transferred in poor condition (e.g. with fin rot) will soon become invaded by bacteria, such as *Vibrio*, because of the cumulative effect of the various stresses; the resulting disease can then spread to apparently healthy stock.

9.2   ACUTE LOSSES

Sudden unexplained losses among apparently healthy fish in seawater are usually due to oxygen starvation or algal blooms. As indicated in Chapter 11, algal blooms are a regular cause of major fish kills on some coasts (Plate 36) and have been known to kill out an entire marine farm overnight. Some marine algal blooms or 'red tides' kill fish by removing oxygen from the water and it may be possible to alleviate the effects by aeration. Where a toxic alga is involved, the fish are killed by the chemical effect on gills or the nervous system and nothing can be done except to harvest marketable fish.

Another very serious form of fish kill is poisoning by jellyfish or 'scouders'. Occasionally scouder populations rise inordinately and may be washed against sea cages by the tide. This may cause them to break up and pieces of toxic nematocyst or sting are ingested by fish and destroy the gill epithelium. Losses from scouder stings can be 50% or more of total stocks (see page 130).

Apart from fish kills, acute losses often signal an infectious disease, such as bacterial septicaemia due to furunculosis or vibriosis. A number of diseases which generally follow a chronic pattern are often preceded by an acute phase and a sudden spate of losses can also be triggered by husbandry stresses acting on a population of fish already weakened by an underlying health problem.

9.2.1   Furunculosis

Formerly, bacterial diseases, especially furunculosis and vibriosis, were the main reasons for heavy losses in marine cage culture. Recent developments, particularly the advent of an effective furunculosis vaccine, and the adoption of the procedures of fallowing of sites, and adoption of single year class sites, have greatly reduced the incidence of these diseases, but furunculosis in particular is still potentially able to cause very great economic losses, particularly to Atlantic salmon crops.

Affected fish usually show a short period of reduced appetite before losses start to occur. Dying fish are often dark in colour and may have haemorrhagic gills, sometimes with congestion around the pectoral and pelvic fins. Internally there is usually blood-stained fluid, the spleen is usually enlarged, rounded and cherry-red and the intestines engorged with blood. If the fish live longer, then muscle lesions (like smaller and less raised freshwater furuncles) may be seen, while small blisters under the skin can produce the occasional raised scale.

Unless vaccinated in freshwater, acute furunculosis is a common problem in smolts soon after transfer to seawater. This can occur because the smolts already had subacute furunculosis at transfer, or because there were some furunculosis carriers in the population but no clinical disease, or alternatively because the smolts were exposed to the disease among adjacent fish populations on entering seawater. Losses may be as high as 30% or more and an outbreak is often followed by recycling of the infection. This means that even after antibiotic treatment has apparently brought the problem under control, another outbreak will occur in a few weeks or months, followed by further subsequent recurrences among the surviving growers. Every effort should be made to avoid introducing the furunculosis organism to a marine site, hence the importance of smolt quality. If carrier fish are detected, and if the smolts have to be used, they should only be transported to saltwater after the bacterial strain they are carrying has been tested for sensitivity to antibiotics. They must then be put onto a diet containing therapeutic levels of that antibiotic over the period of transfer and for 7–10 days thereafter. It is not absolutely certain that even this procedure can guarantee removal of all carriers and vaccination is generally a much more reliable way to ensure non-transfer of infection to seawater.

If a marine farm has a history of furunculosis, it is very difficult to prevent clean smolts becoming infected soon after arrival. The source of infection and continual reinfection is the chronic carrier fish. This is often dark, listless and inappetent, and since it is not receiving any medication provided via the feed it will carry viable bacteria for a considerable time. Such carriers should be eliminated by culling all poor growing or moribund fish. Dead fish probably contain vast quantities of infectious bacteria and *must* be removed daily for hygienic disposal in lime pits or caustic soda.

Even with the advent of reliable vaccines, control of furunculosis will continue to involve a combination of preventive medicine (including the use of furunculosis-free stock), chemotherapy and sensible management, such as keeping apart different populations and year classes wherever possible. The most commonly used drugs in treating epizootics are oxytetracycline, oxolinic acid and amoxycillin. These are normally given in the feed.

9.2.2   Vibriosis

The two most important *Vibrio* species for salmonids are *V. anguillarum*, which is responsible for a wide range of septicaemic outbreaks in stressed fish at any time of year, and *V. salmonicida*, which is particularly prevalent in Norway but occurs elsewhere also. *Vibrio salmonicida* is a serious cause of losses at low water temperatures.

Many species of marine fish are susceptible to vibriosis and many fish can act as *Vibrio* carriers. As with other acute bacterial diseases, salmonids usually show a short period of reduced appetite before losses start to occur. Dying fish are often dark in colour and when opened up they appear very haemorrhagic, with swollen spleens and soft liquefying kidneys (Plates 37(a) and (b)). Fish that survive longer often develop ulcers which erode the muscle. They are usually deeper than furuncles and a large lesion may breach the skin via a small opening through which the highly infective red-tinged exudate is released to infect other fish.

Infection can occur by ingesting infected material from other fish or via skin wounds and loss of scales after grading or transport, etc. predisposes to vibriosis. In practice, *Vibrio* epizootics among salmonids very often can be linked to obvious stresses such as over-stocking, excessive handling or transferring smolts to sea in poor condition. Prompt treatment with antibiotics such as oxytetracycline or oxolinic acid via the feed usually saves those fish that are still feeding. Provided dead fish are promptly removed and husbandry standards are maintained, there need not be subsequent recurring outbreaks as occurs with recycling furunculosis.

Vaccines that give a high degree of protection against vibriosis are now available. They must be given during the freshwater phase of growth and are effective whether given by individual injection or on a mass vaccination basis by bath or spray.

Vibrio ordali is closely related to *V. anguillarum* and *V. salmonicida* and although in most respects the disease it causes is similar, studies on Pacific salmon affected with this organism have suggested that it is responsible for more localized lesions, in heart and body muscle, rather than the generalized septicaemia generally found with *V. anguillarum*.

9.2.3   Hitra disease

Hitra disease or coldwater vibriosis caused by *Vibrio salmonicida*, which was at one time responsible for heavy losses of farmed salmon in Norway, was first seen around the island of Hitra off Norway. It often occurs during the severe winter months and affected fish are anaemic with pale swollen livers. Those fish that do not die quickly, stop growing and show ragged fins, scale loss and skin ulcers. The triggering cause is not known with certainty although it was believed to be related to a combination of husbandry factors and *V. salmonicida* infection. The nutritional status and particularly lipid and vitamin E levels were possibly significant, and now that diets are better formulated and *V. salmonicida* vaccines are available, it is no longer a serious problem.

9.2.4   Pancreas disease (PD)

PD can strike seawater salmon stocks at any time, although it is particularly prevalent in young salmon during their first few months at sea. An outbreak can be explosive, with the virus spreading from cage to cage at great speed and large numbers of fish showing severe clinical signs.

The first sign is that fish go off their food, become lethargic, dark and adopt a strange 'hanging' position in the water. Mortalities can be high in the acute stage, but the main loss is in terms of growth in the large numbers of fish that do eventually return to feed, and in the large numbers of black, thin, wasting fish (postviral myopathy syndrome (PMS), see p. 119). These have to be culled, but prior to culling can act as a serious focus for heavy sea lice infections or furunculosis, which can then extend to the rest of the crop.

There is no treatment for the disease although recent isolation of the causative virus may give hope that a vaccine may eventually be developed. A common observation, however, is that the incidence of PD is greatly reduced in fish populations that have been vaccinated against furunculosis. Presumably, the high levels of inflammatory activity around the pancreas associated with intraperitoneal vaccination with an irritant adjuvant enhances levels of non-specific protectants in the area of the pancreatic tissue.

9.2.5 Infectious salmon anaemia (ISA)

ISA is a serious virus disease that has only been described from Norwegian waters. The virus is spread via movements of infected fish or release of infective materials from processing plants into the sea, so since major controls have been introduced, it appears to be decreasing in significance each year. Affected fish go off the food, become extremely anaemic and die, often after feeding, netting or other minor excitement. The gills are very pale and affected fish are highly susceptible to other conditions.

9.2.6 *Leucocytozoon* infection

Leucocytozoon infection has recently become a serious source of acute mortality in certain stocks of Atlantic salmon cultured in Chile. The cause is an intracellular blood parasite which occurs within the white blood cells. The parasites can be seen in Gram-stained blood smears. There are suggestions that the disease results from a combined infection with the *Leucocytozoon* and a retrovirus, but there is no definite evidence as yet for this.

Affected fish are usually dark and die if stressed by handling. They have a yellow-coloured ventrum, pale gills and the tissues of the abdomen are pale or yellow. There is usually swelling of the tissues behind the eye resulting in popeye. The spleen is usually swollen and the kidney is also occasionally enlarged.

There is no treatment other than minimal stress and careful husbandry. Eventually, a proportion of affected fish recover and are thereafter solidly immune. Since the disease seems to particularly affect certain strains of Atlantic salmon, one way to avoid losses is to avoid use of eggs from these particular strains.

9.2.7 Marine anaemia (plasmacytoid anaemia)

Marine anaemia or plasmacytoid anaemia is a serious cause of acute losses in chinook salmon in British Columbia. It occurs suddenly, and large numbers of growers are suddenly found moribund or dead at the bottom of the cage. Gills of moribund fish are pale and the spleen and kidney, though swollen, are also pale. No infectious agent has been isolated though the possibility of an oncogenic or cancer-causing virus is considered likely.

9.2.8 Rickettsiosis or salmon rickettsial septicaemia

Rickettsiosis is a very serious condition that first occurred in Chilean waters in 1989. Ultimately, up to one-and-a-half million fish succumbed over the period April to December 1989. Coho salmon were originally the sole species affected, but subsequently Atlantic salmon and rainbow trout have been shown to be affected, though not with the seriousness of coho. The causative agent is *Piscirickettsia salmonis*, a strongly Gram-negative, minute bacterium which can only survive and grow within living cells. Thus, it can only be isolated in tissue cultures. *Piscirickettsia salmonis* has also been isolated in Scotland, Norway and Iceland, but mortalities are very low and there may be strain differences.

The disease which occurs shortly after fish are introduced to salt-water can be an acute septicaemia characterized by anaemia and severe swelling of the kidney, with heavy mortalities of fish showing few clinical features other than dullness, inappetence and inactivity. Fish that live longer may develop characteristic raised skin lesions (Plate 38), and focal lesions in the liver (which may be mottled), spleen, heart and other vascular organs. Losses can be 20% or more but surviving fish seem solidly immune to re-infection.

Treatment by incorporation of antibiotics in the food is of limited value since affected fish do not feed, but injection with antibiotics, preferably quinolones, gives therapeutic dose levels which last for 20 days or more which both ensures that all fish are treated and also allows removal of poor fish, during the treatment process, leading to lower densities of better fish.

9.2.9 Amoebic gill disease

Amoebic gill disease was first described in Tasmanian Atlantic salmon farms, but it is now recognized as occurring wherever Atlantic salmon are subjected to high salinity and warm temperatures. The condition is due to invasion of the gill tissues by a para-amoeba which, it would seem, normally lives a commensal existence in the mucus of the gill. Under certain conditions, of which temperature and salinity appear the most significant, it invades the gill tissues in large numbers and excites a very serious host response. This greatly reduces the respiratory efficiency of the gills. Since at high temperatures the capacity of saltwater to hold oxygen in solution is low, such gills are unable to extract sufficient oxygen for respiration, and high mortalities may result if fish are handled or even if low level feeding is attempted.

The best treatment is to reduce the salinity of the water by moving the cages to estuaries or even pumping fresh water over the surface of the cage.

9.2.10   Infectious pancreatic necrosis (IPN)

Previously thought of as a disease of freshwater, it is now clear that IPN can also be a serious cause of losses in salmon in saltwater. The condition may be a recrudescence of acute infection derived from freshwater outbreaks (see Chapter 7, 'IPN') where acute losses have been low, but subsequent stresses have lowered resistance in carriers. More often, IPN losses occur due to exposure of disease-free fish to the virus for the first time shortly after smolt transfer. Often it is the best fish that succumb first, and usually little more than slight abdominal swelling and a slightly swollen vent can be seen. In fish that survive longer, clear or whitish fluid in an otherwise empty gut, possibly small haemorrhages over the pyloric caeca and a swollen vent are found (Plate 39).

Losses may be up to 10% of stock but can be minimized by introducing single-origin fish to fallow sites, ensuring high feeding levels immediately after smolt transfer, removing mortalities and moribund fish promptly, and not mixing fish from certified IPN-free sources with older stocks or stocks from less well-tested origins.

9.3   SKIN PROBLEMS

Skin lesions are a frequent sign of ill-health, and ulcers can develop as a result of bacterial septicaemias, Hitra disease, etc. Traumatic damage due to rough handling or predator attacks can cause areas of scale loss, ulcers or open wounds, sometimes accompanied by damaged eyes. The latter may simply show corneal scarring or progress to total enucleation of the orbit. An attack by certain species of stinging jellyfish will result in characteristic branched, whip-like marks on the skin (Plate 40). However, a number of common diseases of marine salmonids are primarily characterized by skin problems. Most of them are merely of nuisance value and are unlikely to result in heavy losses provided prompt action is taken as appropriate, but one particular problem, parasitic infection with sea lice, has become probably the most serious problem of marine culture in northern Europe.

9.3.1   Parasites

Sea lice (*Lepeophtheirus* or *Caligus*) can be a very serious problem, especially for the cage farmer. One or two lice are expected on wild fish and may actually enhance market value because they indicate the fish has been taken at sea or in the estuary. However, intensive salmon farms provide the ideal environment for massive population expansion of this prolific skin parasite. Affected fish may have a dozen or more parasites, particularly around the head, back or vent area, and suffer severe irritation from the parasite's abrasive feeding. They respond by jumping and by rubbing against the net, often causing ulcers to form which allow the lice to eat further into the flesh (Fig. 9.1). If left unchecked the ulcers will increase, allowing loss of tissue fluids, and hence dehydration, and also offering portals of

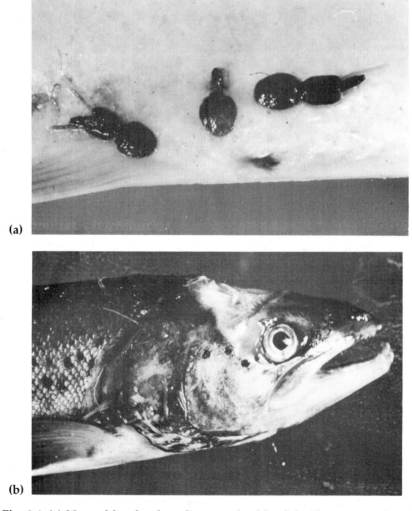

(a)

(b)

Fig. 9.1 (a) Normal levels of sea lice on a healthy fish. They are predominantly located around the vent. (b) Sea lice damage to the head of a caged Atlantic salmon. (*Courtesy of Dr T. Håstein.*)

entry for pathogens such as *Vibrio*. Therefore, the level of sea lice
infestation and damage must be constantly monitored in order that
prompt treatment may be instituted before performance is affected or
losses commence.

There are a variety of possible control measures for lice. Chemotherapy originally involved the use of organophosphorus compounds applied as a bath. Unfortunately, many lice populations are
now resistant to these compounds, but bath treatments involving
hydrogen peroxide or pyrethroids are available. The latter compounds are particularly effective because they kill the early chalimus
larval stages which organophosphorus compounds and hydrogen
peroxide do not (see Chapter 13, Newer bath treatments for sea lice).
Oral treatment methods are now being developed and are likely to
become widely available in the near future. In bath treatments the
cage is surrounded by a tarpaulin and the chemical added. It is
necessary to aerate the fish under treatment and the fish must be
closely observed at all times.

Other non-therapeutic control measures are available which can be
effective in controlling lice numbers. The use of wrasse as cleaner fish
to pick lice off salmon can be effective in some cases. Similarly,
fallowing of sites between crops of fish will help to reduce lice
populations and may prevent any need for treatment of young fish.
If several production sites are being operated within the same sea
loch or fjord it is important that there is integration of their lice
control strategies for maximum effect.

Marine stock can also be infected by ectoparasites such as
Trichodina, *Gyrodactylus* and especially *Costia*. These mostly represent marine species of the parasites, but *Costia* at least may survive
transfer from freshwater with smolts. *Costia* is a particular problem
on the gills of smolts in summer. Prompt diagnosis is essential since
affected fish will lose condition very quickly and mortalities will
occur if the problem is left unchecked. Treatment is by means of
formalin.

Another problem may arise if farms are adjacent to winkle beds,
when the cercaria of the digenetic fluke *Cryptocotyle lingua*, released
in vast numbers from the snails, invade the skin of the fish, producing unsightly black spots.

9.3.2 Myxobacteria

Any form of damage to fish skin is likely to encourage myxobacterial
invasion which will in turn often enlarge and exacerbate the lesion.
This is particularly likely to occur if the fish are overstocked, and
extremities such as the snout are very vulnerable. Another predisposing factor is the onset of sexual maturity since the resulting hormonal
changes make the skin more susceptible to damage. Erosion leads
to sloughing and the consequent ulcers are rapidly invaded by
myxobacteria.

At low temperatures *Flexibacter psychrophila* is able to colonize the intact surface of fins or tail and stimulate considerable thickening of the outer skin surface accompanied by the production of thick tenacious mucus. This coldwater disease affects particularly chinook and coho salmon, which develop greyish-white extremities (Plate 41), and if the skin sloughs, subsequent invasion of the muscles will result in rapid death.

Where the initiating cause cannot be eliminated by better husbandry, transferring affected fish to freshwater will normally destroy the marine myxobacteria involved. In practice there are few sites where this is a real option and, even if it can be done, a number of fish may be lost in the process.

9.3.3   Salmon pox (papillomatosis)

The appearance of greyish, circular, flat warts on any part of the head or body signifies salmon pox. This is thought to be due to a pox virus and usually occurs in freshwater. A small proportion of all smolts from infected hatcheries may be affected. In the sea the situation may be more serious, although the warts do not generally trouble fish until they rub off leaving ulcers which may become secondarily infected (Fig. 9.2). Should this happen during winter, the fishes' ability to heal the lesions will be impaired and myxobacterial invasion may be sufficient to kill the fish (compare this with

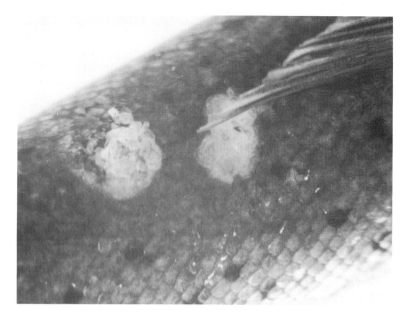

Fig. 9.2 Benign papilloma on the dorsum of marine salmon. It is about to slough.

Plate 39 Smolt with acute IPN. Note the abdominal distension and swelling at the vent as well as the myxobacterial invasion of the dorsal fin.

Plate 40 Jellyfish stings, appearing as whiplash marks on the skin of an Atlantic salmon. Serious damage can be inflicted, especially if the stings affect the gills. (Courtesy of Dugald Campbell.)

Plate 41 Myxobacterial coldwater disease. The larger of the two tails shows the beginning of coldwater disease. The affected part is thicker, greyish, and will quickly become eroded by myxobacteria. (Courtesy of Professor R. D. Wolke.)

Plate 42 Malignant carcinoma following on papillomatosis in an Atlantic salmon. (Courtesy of Dr T. Håstein.)

Plate 43 (a) Typical wasting Atlantic salmon with PMS stage of chronic PD. (b) The pyloric caeca are devoid of fat and very clearly defined in later stages.

(a)

(b)

Plate 44 Nutritional cataract. The lens of the eye of this salmon has a white, dense cataract obscuring the vision and indicating underlying nutritional deficiency or imbalance.

Plate 45 Large granuloma of the kidney around an infection with *Exophiala* mould. (Courtesy of Professor R. H. Richards.)

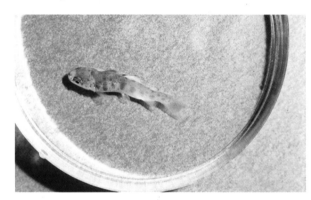

Plate 46 Skeletal malformation. This fish was on an experimental diet deficient in vitamin D, and the spine is consequently twisted.

Plate 47 Fatty liver. The liver is bronze in colour and has rounded edges. There is marked anaemia and much fat around the viscera. (Courtesy of Dr T. Håstein.)

Plate 48 Viscera of a rainbow trout with pansteatitis. The swim bladder is thickened and filled with viscid fluid, and the fish is pale and anaemic.

Plate 49 Gallstones (arrowed) appear as white granular material in the enlarged gall bladder of this Pacific salmon.

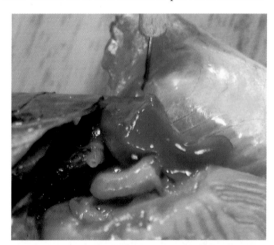

Plate 50 Aflatoxin fed to trout produces this characteristic cancer of the liver known as dietary hepatoma. The liver is very enlarged, soft and often has haemorrhages on it.

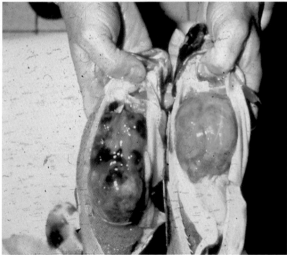

coldwater disease). In Norway it has occasionally been found that salmon pox lesions may become cancerous, and the affected fish will die (Plate 42).

9.4   CHRONIC LOSSES

After recovering from an acute epizootic of furunculosis or vibriosis, it is common for a population of fish to contain chronically diseased carriers, which will be the cause of periodic acute flare-ups on the farm. However, these particular bacterial diseases usually present as acute septicaemic infections, whereas various other bacterial infections are chronic from the start. The latter are typified by fish tuberculosis, which used to be a common cause of continuing chronic losses among farmed salmonids. Nowadays BKD or the various systemic fungal infections are more frequently observed examples of chronic disease. Table 9.1 summarizes the key diagnostic signs of acute and chronic diseases among marine growers, including skin problems described above.

9.4.1   Bacterial kidney disease (BKD)

BKD is typified by sporadic mortalities associated with white nodular lesions in the spleen and kidney. The lesions contain the causative *Renibacterium* organisms surrounded by chronic inflammatory tissue. This usually comes about when symptomless BKD carriers are transferred as smolts to sea and the resulting stress, compounded by changes in kidney function, causes the organism to spread throughout the body. Thus, BKD mortalities are often seen in smolts fairly soon after seawater transfer, but husbandry stress (e.g. dietary changes) can precipitate overt infection in growers also. Such fish cannot be treated but should be removed from the system, in which case the problem often disappears as horizontal spread from a carrier population appears to occur slowly.

9.4.2   Post-viral myopathy syndrome (PMS)

After an acute outbreak of PD virus infection, a proportion of the surviving fish become affected by a chronic muscle disease. The exact pattern varies widely from farm to farm, but the classic signs are thin emaciated fish which usually do not feed or do not digest any feed consumed. These wasting fish have low vitamin E levels, clear fluid in the abdominal cavity and the pyloric caeca are sharply defined. It is necessary to examine the heart and pancreas microscopically to identify the syndrome with certainty as the typical wasting and inflammatory cells are most frequent here (Plate 43). A curious feature of the disease is that after several months a proportion of the affected fish will start to regenerate damaged cardiac muscle, regain their

Table 9.1 Matrix showing the commonest diagnostic signs of disease among growers in seawater

	Losses	Dark colour	Inappetence	Internal bleeding	Wasting	Flashing and jumping	Skin lesions	Abdominal fluid	Gross internal pathology
IPN	++		+						
ISA	++	+	+	+					
PD and PMS	+	+	+		+++			+	
Furunculosis	+++	+	+	++			+	+	
Vibriosis	++	+	+	+				+	
Hitra disease (coldwater disease)	++	+	+	+	++		+	+	
Rickettsia	++	+	++						
BKD	+	+	+		+			+	++
Sea lice	++	+				++	++		
Amoebic gill disease	+	+	+						
Oral ceratathoa	+	+	+		+				
Fading smolt syndrome	+	+	+		+				
Systemic fungal infection	+				+			+	+++

pancreatic function and start feeding and growing again. However, a significant number will need to be culled.

9.4.3 Stress-mediated infectious pancreatic necrosis (IPN)

Where a proportion of fish has suffered IPN, even in a mild form, transfer from the hatchery to saltwater can result in persistent mortalities with a variety of clinical signs. In addition to pancreatic damage evident in the laboratory, there is usually bacterial involvement often causing skin lesions, meningitis with resultant nervous signs, or generalized septicaemia. It is difficult to know how much emphasis to place on the virus and to what extent the bacteria are acting as primary pathogens. However, it is a feature of such infections that losses appear stress-related, antibiotic therapy only alleviates for a short time and the position may sometimes only improve if new virus-free stock are introduced after culling of infected stocks and a period of disinfection.

9.4.4 Tuberculosis (TB) and nocardiosis

TB and nocardiosis are two very similar chronic infections which result in poor growth and the development of granular, lumpy lesions throughout the body. Often the only outward sign is emaciation and darkening of the skin, but at post-mortem examination the white lesions are easily seen in the liver, spleen and kidney. The usual mode of infection is feeding unprocessed guts from fish that had themselves suffered from the disease. If the disease is suspected, confirmation is by identification of the causative organism under the microscope. There is no satisfactory treatment and control is by avoiding the use of infected fish offals as a dietary source.

9.4.5 Cataracts

A common clinical sign of ill-thrift in growers in seawater is opacity of the lens known as cataract affecting one or both eyes. This can be associated with physical abrasion, or be a nonspecific indicator of many chronic disease processes, or be due to poor acclimatisation in smolts newly transferred to sea.

Occasionally cataract may be associated with marginal nutritional deficiency of essential dietary components, such as zinc or vitamins A or E. In such conditions, rapidly growing fish can develop deficiencies manifested in sudden loss of growth and the development of bilateral cataracts in the lens of the eye. Such cataracts are very different from those caused by, for example, the eye fluke, and are generally indicative of a more generalized metabolic fault. Affected fish have spots or generalized whitening of the lens of the eye, readily seen in bright light (Plate 44). They are usually darker in colour, have a reduced appetite, and their growth is distinctly retarded even though they do not necessarily appear blind and may still feed. Often the condition only affects the fastest growing fish in a population, and may only affect fish fed on one specific batch of diet. Once the growth rate has been retarded, fish may start to thrive again, but rarely do the cataracts resolve and affected fish do not make up what they have lost. Often defining the exact cause of cataracts is very difficult.

9.4.6 Systemic fungal infection (mycosis)

As with TB, if fish offals containing the spores of *Ichthyophonus* fungus are fed to salmonids without prior pasteurization, the fungus can invade the gut and then spread progressively throughout the internal organs (see Chapter 10). The result is a chronic wasting disease involving sporadic mortalities of fish showing obvious internal pathology. Such systemic fungal infections are not uncommon in farmed salmonids, even where only compounded diets are fed, but here, instead of *Ichthyophonus*, the fungal species usually encountered include *Exophiala*, *Phialophora* and other members of the black moulds, which occur commonly in soil and presumably contaminate the feed during processing and storage or possibly enter the fish independent of the diet (e.g. by swallowing water heavily contaminated by spores, particularly at the hatchery stage). Affected fish have large growths, containing the fungus, in the kidney or spleen (Plate 45). No treatment is possible and all feedstuffs should be packed and stored in dry condition and fed soon after manufacture to avoid diet-related problems of this kind as well as the allied risk of mycotoxin production. An occasional problem which superficially resembles systemic mycosis is fibrosarcoma of the swim-bladder which grossly distorts the kidney and swim bladder region. It is an invasive cancer which may be caused by a virus and will eventually kill affected fish.

Chapter 10
Nutritional Diseases

With the almost complete switch from trash fish or offal diets towards use of scientifically formulated compound feeds, diseases due to malnutrition, i.e. incorrect dietary composition, are becoming uncommon. It is, of course, easy to underfeed through inexperience, which may well lead to losses from cannibalism or secondary disease problems, and this is most likely to happen at first feeding. Atlantic salmon fry in particular may not come onto feed at all unless food of the appropriate particle size is introduced at the correct time and water flow rates are such as to bring the feed particles down to the fish lying on the tank bottom. Apart from such management problems, the occasional nutritional deficiency will usually show up most obviously during the first few weeks of feeding as fry use up their natural reserves of micronutrients. For this reason many hatcheries use more than one brand of starter feed, often in combination, whereas deficiency syndromes and malnutrition in growers are probably so rare and non-specific as to make this precaution unnecessary.

With the exception of certain problems arising from toxin containing fish diets (notably botulism/bankruptcy disease), nutritional diseases tend to be chronic conditions which appear over an extended period of time. The usual signs comprise low-level mortalities, wasting, reduced conversion efficiency and internal changes, particularly in the liver, kidney and skeleton (Plate 46). Because of this often ill-defined picture of poor health and their comparative rarity, nutritional diseases will be divided up according to the nature of the dietary component involved, rather than the outward appearance of the fish. Four categories of nutritional diseases will be considered; the first three arise from feeding levels of vitamins, minerals and trace elements, and fats which are outside the required prosperity range for salmonids. The fourth category comprises miscellaneous conditions which are thought to have a nutritional origin, including those arising from the presence of contaminants in the diet.

10.1 VITAMIN DISEASES

Vitamins are usually divided into those that are soluble in water (the B group) and those that are soluble in fat (all the rest). Deficiencies of most vitamins in each class have been found to cause disease in salmonids at some time or another, but B, C and E are the most common deficiencies.

10.1.1   Fat soluble vitamin deficiencies

Deficiencies of vitamins A, D, E and K can probably all cause diseases of salmonids, with poor growth and anaemia being common symptoms. Vitamin E is required for fat digestion and for muscle development and there may be insufficient of this vitamin to deal with diets having a high fat content (see 'Fat diseases'). The main distinguishing features of deficiencies of vitamins A, D (and C), and K are blindness, bone malformations and impaired blood coagulation, respectively.

10.1.2   Water soluble vitamin deficiencies

Thiamine (B₁) deficiency
Thiamine (B₁) deficiency most commonly arose due to the use of certain trash fish diets, e.g. herring, which contain an enzyme (thiaminase) that destroys thiamine. This results in brain degeneration with consequent loss of balance, nervous convulsions and blindness. In countries where trash fish feeding is commonly practised, it is usual to incorporate an additive into the diet containing high levels of thiamine in order to prevent the deficiency.

Riboflavin (B₂) deficiency
Eye changes are characteristic of riboflavin deficiency, notably lens opacity (cataract) and haemorrhages into the eye. Affected fish are usually dark in colour.

Pantothenic acid deficiency
Deficiency of pantothenic acid causes an uncommon but characteristic gill disease in fast growing salmonid fry. These show loss of appetite and gasping, due to the gill lamellae thickening and then becoming fused to each other. This nutritional gill disease can be distinguished from more common gill problems, e.g. bacterial gill disease, by the microscopic appearance of the fused lamellae.

Other B group vitamin deficiencies
Pyridoxine deficiency can cause similar nervous signs to those of thiamine (B₁) deficiency, with the occurrence of convulsive fits. Niacin deficiency may cause uncoordinated muscle movements and swollen gills. Biotin deficiency may affect the skin of salmonids and render it more liable to parasitic infections. These and other deficiency syndromes (e.g. vitamin C) are usually characterized by reduced appetite and growth and a tendency to anaemia.

10.2 MINERAL AND TRACE ELEMENT DISEASES

A number of elements are considered essential for salmonids including calcium, chlorine, cobalt, copper, iodine, iron, magnesium, manganese, phosphorus, potassium, selenium, sodium, sulphur and zinc. Deficiency of these elements is rarely a problem, although iodine deficiency used to cause swelling of the thyroid gland (goitre) prior to the use of fishmeal in compounded diets. 'Induced' deficiencies can sometimes occur if a high level of one mineral, such as calcium, interferes with the uptake of other essential trace elements present in the diet, such as manganese or zinc. Thus, it has been shown that diets containing high levels of ash, and hence calcium, need to be supplemented with additional zinc to avoid the risk of induced zinc deficiency causing cataracts and blindness in salmon or trout.

In fact the requirement for certain of these elements is very small and the presence of trace elements in larger amounts can readily cause disease. This can be a particular problem with heavy metals such as iron, copper and zinc, which may enter a water supply from natural rock deposits, metal pipes or with industrial effluents. Iron salts can form a precipitate on fish gills or on the surface of eggs causing asphyxiation of the embryo. Copper and zinc toxicity may be manifested by respiratory difficulties, with pale gills becoming coated with mucus. The toxicity of these heavy metals is often markedly increased in soft water.

10.3 FAT DISEASES

Salmonids have specific requirements for certain essential fatty acids, but deficiencies are unlikely to occur under farm conditions unless rancid diets are fed. The lipid component of the optimum fish diet contains high levels of unsaturated fats. These are particularly easily destroyed by oxidation which makes them rancid and unusable by the fish. Vitamin E and other specific antioxidants are usually added to compound feeds to protect against this, but if the feed is stored in warm, moist conditions these antioxidants can become exhausted. Fish fed on such diets are likely to develop lipoid liver degeneration due to fatty infiltration of the liver, giving a typically bronze-coloured liver which in mild cases is only discovered at gutting (Plate 47). However, in more severe cases the liver swells up and muscular degenerative changes occur, as well as anaemia resulting from reduced blood formation. Such fish may show haemorrhages in the viscera, are weak and anaemic looking and may die when handled.

Two other conditions that are believed to be associated with disordered fat metabolism are pansteatitis and gallstones. Pansteatitis is a condition where the fat in all parts of the body becomes inflamed, and often the swim bladder becomes swollen and filled with viscid fluid, the fish becomes anaemic and the muscle degenerates, becoming very soft (Plate 48). It is thought that pansteatitis is associated with the incorporation of unsuitable fishmeals into a diet.

Gallstones (Plate 49) is the term given to the accumulation of hard calcified material in the gall bladder of the liver. The exact cause is not known, but it is often associated with certain types of lipid in the diet.

10.4   DIETARY INFECTIONS

The use of diets containing micro-organisms pathogenic to salmonids can allow infection to become established within fish farms. Although usually associated with the feeding of trash fish diets, even compounded diets can become infected (e.g. with moulds) and cause disease.

10.4.1   Tuberculosis (TB)

The use of moist diets made from trash fish infected with Mycobacteria can cause TB in farmed salmonids. This is a chronic wasting condition resulting in poor growth and losses in fish, which develop granular and lumpy lesions throughout the body. Often the only outward signs are emaciation and dark coloration, but at post-mortem examination the whitish lesions are easily visible, particularly in the liver, spleen and kidney. Laboratory diagnosis is needed to differentiate TB from BKD and internal fungal infections (see below). There is no satisfactory treatment and the disease can only be prevented with certainty by prior pasteurization or ensilage of trash fish diets.

10.4.2   *Ichthyophonus*

Ichthyophonus is a similar disease to TB although the causative agent is a fungus rather than a bacterium. When trash fish which are infected with *Ichthyophonus* are fed to salmonids, the fungus invades the gut and then spreads progressively throughout the internal organs. Affected fish fail to grow and, when opened at post-mortem examination, whitish nodules are evident, particularly in the heart, muscles, gut, liver and kidney. Diagnosis is made by squeezing one of the white lesions onto a slide; the characteristic branching structure of the fungus is evident under the microscope. Again, treatment is not possible and prevention involves pasteurization or ensilage of infected trash fish.

10.4.3   Miscellaneous fungal infections

The use of compounded diets heavily contaminated by various species of fungal spores can result in a similar picture to *Ichthyophonus* with low-grade recurring losses showing severe visceral infection at post-mortem. Various members of the *Fungi imperfecti* are usually involved which occur commonly in the soil, and it may be that some infections occur initially in hatchery tanks via contaminated water rather than via the diet. Nonetheless, it is important to remember that mouldy feed is a potent source of infection (or even toxins: see below), hence packaging and dry storage of feed are of paramount importance.

10.5   DIETARY TOXINS

Occasionally fish diets have become contaminated by toxic materials, such as weedkillers, which can cause high losses in fish consuming such diets. With modern quality control of raw materials, it is unlikely that dietary toxins will be despatched from feed mills. However, the fish farmer should be aware that compound feeds invariably contain mould spores which will start to grow in suitable moist surroundings, e.g. broken bags. Some such moulds, e.g. *Penicillium*, are well known as producing mycotoxins which cause poor growth in conventional livestock and can have the same effect when fed to salmon or trout. If feed with a high level of mycotoxins is used, large losses can occur and affected fish show haemorrhages in the viscera, sometimes accompanied by a membranous tissue in the abdominal cavity. The fish farmer should aim to avoid such problems by suitable feed storage and should be aware of two other dietary toxins that can occur, as follows.

10.5.1   Aflatoxins

The presence of large tumours in the liver, particularly in rainbow trout, is usually indicative of aflatoxin poisoning (Plate 50). When groundnuts and cottonseed are stored in warm, damp conditions they sometimes become mouldy, and aflatoxins are poisons produced by these moulds. Some months after trout are fed even minute quantities of mouldy oilseeds, liver tumours (hepatomas) appear and these can spread rapidly through the body and cause large losses. Cottonseed can also cause disease more directly due to the presence of a toxic pigment causing inappetance and fatty deposits in the liver and kidney.

10.5.2   Botulism

Botulism is a disease of many animals including fish and also humans in which it is nearly always rapidly fatal. It arises due to the ingestion of the toxin produced by a bacterium, *Clostridium botulinum*. This organism is usually in harmless spore form, but under certain conditions of warmth and oxygen lack, it may start to produce the lethal *botulinum* toxin.

Botulism can occur in salmonids either because wet trash fish diets have gone rotten before being fed or due to fish cannibalising trout carcases rotting on the bottom of a pond. Affected fish consuming the toxin suddenly go off their feed and develop nervous signs, sinking to the bottom of the pond apparently lifeless and then twitching back to the surface before losing station and dying in a matter of hours with no obvious pathological signs.

The spores of the *botulinum* organism occur commonly in soil and mud and can therefore be found sometimes in the gut of healthy salmonids, particularly trout in earthen ponds. Since putrefaction is necessary for the toxin to develop, it is vital to remove any dead fish from ponds promptly for hygienic disposal.

When operating a smoking kiln for processing fish, care must be taken to ensure that fish are hygienically gutted and that the conditions under which toxin formation can take place are avoided. Otherwise there is a remote possibility that fatal human botulism can occur from eating the contaminated product, particularly if it has not been pre-cooked.

Chapter 11
Fish Kills

If a sudden mass mortality occurs over a period of hours (or less) among fish that have been behaving and feeding normally, then the situation is designated a fish kill. Such an event cannot be due to infectious disease and may involve the possibility of litigation. Therefore, it is vital to act swiftly, not only for the purpose of saving as many fish as possible, but also to ensure that all available information is conserved.

Fish kills can be caused in a variety of ways, but the commonest is lack of oxygen. This often occurs in freshwater under conditions of high temperature and low water flows when the fish are fed heavily. In trout ponds or cages with much algal growth, oxygen deficiency may build up in the early hours of the morning, particularly if there is no wind to mix the various water layers (this is because plants use up oxygen during the hours of darkness). In such cases the kill may not be total and fish may be seen gasping in obvious distress. Blocked filters or pump breakdowns usually precipitate heavy fish kills unless noticed immediately, and the dead fish may show haemorrhages in the skin and within the gut cavity.

Many forms of pollution kill fish by lack of oxygen. Thus, most agricultural effluents, e.g. silage liquor, decrease the oxygen-carrying capacity of the water until there is insufficient oxygen available for fish to survive. Whenever pumps are closed down for maintenance, care should be taken to ensure that the consequent rotting of plants and organisms within the pipelines does not result in lethal pollution when the system is put into operation once again.

Cyanide kills by preventing fish from utilizing oxygen, and poachers frequently use various preparations of cyanide. Such fish will appear to have died quietly without any obvious signs, although they may possess a characteristic 'burnt almonds' smell. Many industrial effluents are toxic to fish, particularly heavy metals and hydrocarbons. Lead, copper, iron, zinc and molybdenum have all been implicated in fish kills. Great care should always be taken when handling disinfectants on fish farms. Iodophors, which are so useful for killing fish viruses, must not be allowed to come into contact with live fish; if they enter a fish pond accidentally, the fish will sometimes literally leap out of the water. Insecticides and weedkillers are often a problem and most organochloride and organophosphorous compounds are highly toxic. In the case of dieldrin poisoning, the gills may appear yellow in colour.

129

In areas subjected to deposits of acid rain, that is regions where particular geological characteristics result in the soil and water having very low buffering capacity, and where coniferous forestry is often practised on a large scale, the pH of the water may fall well below pH 5 and its solubility for heavy metals such as aluminium and manganese increases significantly. Sudden spates will result in great unease in the fish stocks, gill function is reduced and any fish that have just been fed are likely to die of oxygen deficiency. If high levels of dissolved metals and low pH persist, the gills suffer chronic damage and in some cases fish hatcheries may have to close down. The only long-term means of control is the provision of an automatic lime dosing system which supplies buffering alkali to the water supply as required.

Probably the most serious threat faced by the marine farmer is the risk of a fish kill resulting from a marine algal bloom. The marine plankton consists of a very wide range of minute animals and plants which increase and decrease cyclically according to the season. Occasionally, particularly in the midsummer months when sunlight hours allow maximal photosynthesis, one or more species of phytoplankton, or marine algae, may undergo a massive increase in numbers, which is defined as an algal bloom.

Algal blooms can turn whole seas red, orange or other colours depending on the identity of the organism responsible. The effect they have on fish also varies for the same reasons. Some produce nerve poisons or gill poisons, others are sharp and irritant, or merely remove all of the oxygen from the water when the bloom collapses. The only feasible procedure, other than immediate harvesting or towing the cages to a more open site offshore, is to provide maximum aeration for as long as the bloom is observed in the area. Even use of a powerful outboard motor adjacent to the cages may help to reduce mortalities. Algal toxins have also been reported as killing salmonids in freshwater lakes during warm weather, although this may again be due in part to removal of oxygen by the algal blooms at night. Finally, it should be borne in mind that most chemicals added to the water for treating fish diseases are themselves toxic to fish at all except very low concentrations.

With the advent of large offshore cages capable of supporting heavy net cages in strong currents, large-scale jellyfish or scouder attacks have begun to become a serious problem. The jellyfish may bloom in such large numbers and be deposited against the net by the tide in such a way that they block the passage of water currents and cause anoxia. Much more common, however, is the release, from jellyfish broken up against cage barriers, of long strands of poisonous nematocysts. The commonest cause of such problems is the lion's mane jellyfish, *Cyanea capillata*, which in certain bloom years may occur in very large numbers in coastal waters. When sudden entrapment of larger numbers of jellyfish against nets occurs, usually in summer, then catastrophic losses can occur. Up to 200t of Atlantic salmon have been lost over a few days, with chronic damage to many more.

Fish are affected in two ways. The pain and irritation of stings across the head, eyes or flanks can cause them to rub against ropes or nets to the extent that they can actually enucleate the eye, or cause 2–3 cm haemorrhagic ulcers on the head (Plate 51(a)). These rapidly become secondarily infected with bacteria. More seriously, if strands of nematocysts are passed over the gills they induce a very severe response in the gill in the form of mucus production and cellular proliferation so severe that if the fish has not died almost immediately from toxic shock, it succumbs within a few hours from respiratory failure (Plate 51(b)).

No treatment is possible, only prevention by removal of jellyfish intact as they approach predator nets. Once a heavy strike has taken place, fish should be harvested as soon as possible after they become active again (signifying degeneration of ingested toxin), as they are unlikely to be able to be fattened further, and secondary infection of the damaged skin will almost certainly lead to a major bacterial problem in fish, which will be difficult to treat as they rarely feed well afterwards.

On discovering a fish kill, it is advisable to call the police immediately unless the cause is obviously the farm's own fault, e.g. a blocked filter, or an act of nature, e.g. an algal bloom, as all such losses should be investigated for malice and an independent report is invaluable if civil action or insurance claims are involved. The manner of dying and the symptoms shown should be written down and the extent of the kill throughout the farm should be noted. Think before you act! If an affected pond drains into a stocked outlet channel, it may be preferable to shut it off rather than increase the flow rate and kill the fish in the channel as well. Nevertheless, the prompt use of aerators, auxiliary pumps, etc. and transferring survivors to clear water can often reduce losses considerably.

Water samples should be placed in clean (lead-free) bottles and sealed and labelled in the presence of two witnesses. It is also helpful to place at least one water sample in the freezer. Newly dead fish should be placed in deep freeze. Dying fish should be killed, and small cubes of the organs (especially the liver, kidney, gut, skin and gills) should be placed in a pickling liquid – 1:10 solution of 40% formalin, or buffered formal saline. If cyanide poisoning is suspected, heads of affected fish should be forwarded to the appropriate laboratory. Photographs may be a useful aid should litigation ensue, and a comprehensive account of the incident should be recorded and forwarded to the appropriate bodies if any further action is envisaged.

Chapter 12
Diseases of Wild Fish and Broodstock

Diseases of wild fish have an intrinsic interest for anglers or netsmen because they can affect the abundance and aesthetic appeal of the quarry. They are also extremely significant to the fish farmer since wild fish can act as carriers for a number of the major disease agents. In addition, many fish farms rely on the capture of mature wild fish to provide the eggs and milt for hatchery operations. Broodfish on the farm are usually kept in somewhat similar conditions to those in the wild, being less densely stocked than growers. They share many disease problems with wild fish and these will be considered together in this chapter.

As in the farm situation, very severe losses are invariably due to non-infectious causes, and are considered under 'Fish Kills' (Chapter 11). In heavily stocked fisheries with abundant aquatic vegetation, fish kills due to oxygen lack are a not uncommon occurrence. Pollution is always a danger and, in the case of valuable adult fish, malicious use of substances such as cyanide, carbide or dynamite must be considered.

Leaving aside fish kills, diseases of wild fish and broodstock can be broadly divided into three groups: acute infectious diseases; diseases recognized mainly by skin lesions; and a miscellaneous group recognized incidentally at post-mortem. Except where relevant to management of broodstock, details of therapy are omitted since this is not generally feasible for wild fish. For those diseases shared by wild and farmed fish, such information is available elsewhere in the text.

12.1 ACUTE INFECTIOUS DISEASES

Acute infections occurring among broodstock or in the wild are generally due to bacteria. Broodstock show a short period of inappetance before the onset of quite heavy losses, which are usually the first sign to the farmer of any disease problem. Under certain circumstances, acute infections will also result in the appearance of haemorrhagic ulcers on the outside of the fish, but these will be described in the next section.

If dead or dying fish with no external signs are opened up, acute infections can be recognized by a variety of internal changes, the commonest of which is the presence of haemorrhages on the internal organs. During high water temperatures, this is the typical picture shown by young salmonids suffering from acute furunculosis due to *Aeromonas salmonicida* infection.

If small white lesions are also visible scattered throughout the spleen, kidney and liver, then the causative organisms may be *Renibacteria*, which can be responsible for sudden losses in some areas. The species most severely affected is the Atlantic salmon and the condition is so frequent in certain years in the Scottish river Dee that it is referred to as Dee disease. Unlike furunculosis, Dee disease can also occur at very low temperatures (about 4°C). In this case, diseased fish show a white tenacious membrane stretched over the liver, kidney and spleen, instead of the combination of white spots and haemorrhage shown at high temperatures.

The bacteria that cause these infections often survive between outbreaks of disease within the tissues of carrier fish. These fish can release them into the water to continue the infection while remaining apparently healthy. Because of this danger, eggs from rivers with a high incidence of such diseases should not be taken for stocking fish hatcheries. For the same reason it is important that any seriously affected fish are removed from the river and buried.

12.2   DISEASES CHARACTERIZED BY SKIN LESIONS

A variety of diseases of wild fish and broodstock are manifested by obvious skin abnormalities. These may be conveniently classified into those diseases associated with sexual maturity, and others, and they will be discussed accordingly in this section. In general, salmonids showing the pale skin patches associated with fungus infection may be assumed to be suffering from one of the diseases associated with sexual maturity.

12.2.1   Diseases associated with sexual maturity

The hormones that cause the gonads to develop within salmonids bring about a variety of additional effects. Thus, in the case of Pacific salmon they result in the virtual breakdown of all bodily systems and death ensues soon after spawning. In most salmonids the stomach and intestine change markedly at the onset of sexual maturity and become incapable of digestion. The skin also changes in various ways depending on the species. It actually becomes thicker in male fish and the mucus alters to a more viscous consistency.

Salmonid skin appears more prone to infection while undergoing these changes around spawning time. Physical damage to the skin is also more liable to occur while fish are migrating to their spawning redds, and fighting among broodstock males often results in injuries. Such wounds heal more slowly on account of the low water tempera-

(a)

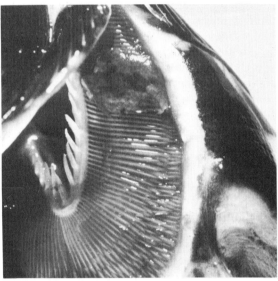

(b)

Fig. 12.1 Fungus infection. (a) *Saprolegnia* covers most of the body in this spawning brown trout. (b) *Saprolegnia* infection of the gill is particularly serious and death usually follows rapidly.

ture. In addition, there may be a greater number of fungal spores present in rivers during this season. Certainly many fish become severely infected with fungi, especially *Saprolegnia*. Creamy-coloured patches of fungus appear on any part of the skin and these often kill the fish outright (Fig. 12.1). They may follow skin trauma, e.g. due to seal damage, stab wounds caused by birds, or netting marks, but can also result from ulcers caused by disease. If lightly affected fish regain the sea, the fungus usually disappears, although contraction of the resulting scar can sometimes cause permanent deformity.

Fish that are severely affected with fungus should be removed from the river, killed and examined, before being buried, preferably in a lime pit. The presence of ulcers on the head or haemorrhages inside the body should suggest the possibility of two specific diseases, which are now discussed in more detail.

Autumn aeromonad disease

Autumn aeromonad disease is a condition that mainly affects adult brown trout during the spawning season. Such fish show patches of skin fungus due to *Saprolegnia* and sometimes the skin overlying the gut cavity may show severe inflammation (Plate 52). The main feature of the disease is the state of the internal organs which are usually very haemorrhagic. The kidney is often completely liquified so that when incised it flows out of its capsule.

The disease is associated with generalized bacterial infection, usually with *Aeromonas hydrophila* or related bacteria, which can be isolated from all organs in the laboratory. However, the primary cause would appear to be the stresses associated with spawning. Thus, immature fish, or adults that are not spawning, may carry the bacteria responsible for the disease, but are themselves completely unaffected.

Ulcerative dermal necrosis (UDN)

Ulcerative dermal necrosis (UDN) is a disease of unknown cause which was originally described as affecting Atlantic salmon and sea trout as they came into freshwater. UDN is confined to adult fish and has at present only been described as occurring in the UK, Republic of Ireland, France and Sweden.

In migratory salmonids the disease has occurred in epizootic levels three times in the past century. Characteristically it occurs during the colder months of the year. It usually starts as the fish enter freshwater for their spawning run, and the first signs are very small greyish lesions on the side of the opercula, above the eyes or on the snout (Plate 53). These either heal or spread to involve a greater area of the head. They then ulcerate to produce a reddish wound, which becomes infected by bacteria or fungus from the surface of the fish or the water. The fungus spreads from the initial site of infection to the rest of the weakened fish, which thus becomes indistinguishable from any other fish with a severe fungus infection.

There is very little information available on the disease in brown and rainbow trout. Certainly a disease similar to classical UDN of salmon occurs in maturing fish in certain fish farms and rivers during autumn. The lesions are confined to the head and are not at first affected by fungus. In the absence of any diagnostic test only tentative diagnosis is possible. This is usually based on the findings in the early stages of uninfected head lesions in adult fish.

There is no evidence that lightly affected fish cannot spawn successfully or that their offspring are either more or less susceptible to the disease. However, in view of the possible infectious nature of UDN, it would seem sensible to take hygienic precautions with fishing tackle and clothing if moving from an infected to an uninfected water, and to avoid the introduction of eggs from an infected river to a previously clean hatchery. If lightly affected fish are taken for stripping, the disease can usually be cured by treatment with malachite green, since the internal organs are not affected until secondary infection has progressed.

12.2.2 Diseases unassociated with sexual maturity

Several diseases which are sometimes seen among wild salmonids are not specifically linked to sexual maturity. These include skin tumours and certain acute infections characterized by skin ulcers, as well as the IHN virus which takes an annual toll of wild fry in western US rivers, and VHS and whirling disease which are now serious problems in north-western US rivers.

The presence of skin lesions on dead and dying fish without visible signs of fungus infection is usually evidence of acute bacterial infections. Grey head lesions, which may superficially resemble UDN ulcers, are usually caused by *Flexibacter columnaris* if occurring at temperatures exceeding 20°C. This columnaris disease is mainly found in wild Pacific salmon and is rapidly fatal over 25°C. Microscopic examination of a scraping from ulcers of affected fish demonstrates the characteristic columns of bacteria and permits differentiation from UDN.

A more common skin manifestation of acute bacterial disease is the appearance of furuncles and/or deep haemorrhagic ulcers. Furunculosis, due to *Aeromonas salmonicida* infection, is usually evidenced in older fish by the classical sign of furuncles developing over the sides and back (frontispiece (d)). These sometimes burst and release haemorrhagic and highly infectious fluid. *Vibrio anguillarum* infection is a major cause of losses among salmonids returning from the sea into estuarine conditions. Vibriosis results in the formation of deep haemorrhagic abscesses within the muscles, with concurrent darkening or ulceration of the overlying skin. When fish affected with either furunculosis or vibriosis are opened up, haemorrhages of the internal organs are usually found, sometimes accompanied by swelling of the spleen and liquefaction of the kidney. Although affected broodstock may sometimes be saved by antibiotic and sulphonamide treatment, affected wild stock are best killed and buried and every practical precaution taken to limit the spread of these contagious diseases.

Tumours are not infrequently encountered among wild fish, including both benign and malignant tumours. A malignant tumour (cancer) grows rapidly and spreads throughout the body, whereas a benign tumour merely increases in size slowly at its original site. Internal tumours sometimes occur, particularly arising from blood-forming tissues. These are usually seen only on gutting the fish, except in the case of tumours of the thymus gland beneath the gills. Several such tumours have been reported and may be recognized by the swollen head appearance of affected fish resulting from pressure on the opercula (Fig. 12.2). Skin tumours usually comprise whitish-

Fig. 12.2 Thymic tumour in a rainbow trout. This tumour appears as a large pink swelling at the top of the gill cavity, and consists of altered tissue from the thymus gland. *(Courtesy of Mr J.F. McArdle.)*

coloured lumps or raised, bleeding ulcers, especially if malignant (Plate 54). A fairly common benign tumour of salmon skin is probably caused by a virus. This is known as salmon pox and results in warty growths on the fins and body. Salmon pox usually clears up eventually, but if warts become rubbed off the fish, the resulting ulcers are readily infected by fungi or bacteria.

Non-cancerous growths, such as cysts, are not uncommon in salmonids (Plates 55 and 56), but these are usually restricted to one or two individual fish and not usually of any great importance.

12.3 DISEASES DISCOVERED AT POST-MORTEM

There are a number of parasitic conditions that affect wild salmonids without usually causing frank disease. These are often discovered incidentally when the fish is captured, particularly when cleaned prior to cooking. Most of these parasites are incapable of developing and causing disease in man, even if eaten uncooked. However, they are often unsightly and reduce the aesthetic appeal of the catch.

12.3.1 External parasites

Eyes
Blindness due to eye fluke infestation can be a severe problem in certain lakes where there are a large number of infected snails present.

Gills

Monogenetic flukes may be just visible on the gills of salmonids held in heavily stocked fisheries. After adult salmon have returned to freshwater they can also develop heavy infections with gill maggots. Since these maggots can survive on kelts returning to the sea, freshly run repeat spawners may be readily recognized by the very large numbers of maggots sometimes present (Fig. 12.3). Gill maggots are not normally considered harmful to salmon, but secondary fungal infection of damaged gills may cause occasional losses among kelts.

Skin

Skin infections with small Protozoa and flukes can cause irritation and excess mucus secretion when the causative organism is often invisible to the naked eye. White spot (Ich) can be a considerable problem in young fish under conditions of high water temperature. This is especially so when the water is shared by a population of coarse fish and there is a low flow rate.

Of the larger crustacean parasites, both fish lice and sea lice are found on salmonids. The fish louse is found only in warmer waters and is of little significance except when heavy infestations provide sites for secondary infection. The presence of sea lice on wild salmon is welcome evidence of freshly run fish and they are usually sited near the vent or in the tail region.

12.3.2 Internal parasites

The only salmonid parasite that is very dangerous to man is a small white coiled worm occasionally seen on the surface of the liver or

Fig. 12.3 Parasitized gills. The gills of this Atlantic salmon are heavily infected with gill maggots (see Fig. 4.11), which results in a coat of thick mucus and small haemorrhages.

other internal organs of salmon and sea trout (Fig. 12.4). These *Anisakis* worms are parasites of marine fish, and salmon can become infected by eating krill. *Aniskakis* can cause severe intestinal damage to humans if swallowed live, but this is unlikely to happen unless raw fish is eaten.

Another parasite that is pathogenic to man is the tapeworm, *Diphyllobothrium latum* (Plate 33). This is one of a family of tape-worms which·use various fish as host for one of their intermediate stages. The other more common species usually form an adult tape-worm in the gut of fish-eating birds. In heavy infections, the larval plerocercoids can cause mortality of fish as they migrate through the body tissues to the abdominal cavity or muscles. If infected fish are eaten uncooked, e.g. as lightly salted fish, the adult *D. latum* tape-worm can survive within the human intestine, where it has been known to cause pernicious anaemia due to induced vitamin B_{12} defi-ciency. *D. latum* is only common in rural fish-eating communities.

Much more common are the red coiled filariid or spilurid worms which move across the body surface of heavily infected fish when they are cut open. Although of no danger to man, these worms are particularly unattractive. Salmonids usually respond to their pres-ence by laying down a large fluid-filled cyst within the abdominal cavity. This usually represents an intermediate larval stage of the parasite, whose adult life is spent in the stomach of fish-eating birds.

Sometimes when the angler cuts into the flesh of apparently healthy salmon or sea trout, certain areas of muscle may appear to have been replaced by a yellow creamy fluid. This is the condition of milky flesh disease, due to a protozoan parasite, *Henneguya*, and severely affected fish are useless for the table.

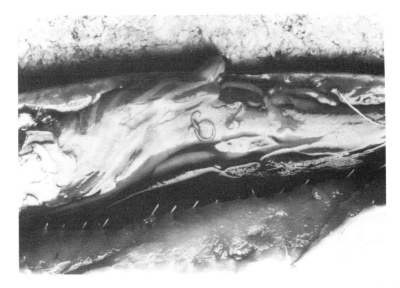

Fig. 12.4 *Anisakis* worms on the viscera of Atlantic salmon. (*Courtesy of Dr R. Wootten.*)

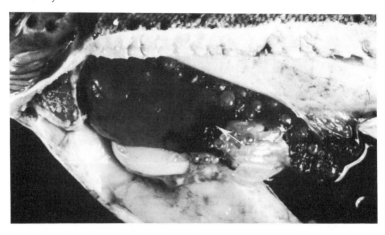

Fig. 12.5 Broodfish collapse showing oedematous abdominal viscera and small haemorrhage (arrowed). (*Courtesy of Dr H.W. Ferguson.*)

When gutting salmon and trout, certain other parasites may be seen, particularly if the gut is accidentally incised. Thus, if the swim bladder appears to be stuffed with cottonwool, this is probably due to large numbers of *Cystidicola* worms, which are of little significance. More harmful to fish are certain thorny-headed worms (*Acanthocephalus*) present in wild salmonids in most areas. They become attached to the intestinal wall (Plate 57) and in severe infections can cause obvious gut haemorrhage and loss of weight, increasing the risk of secondary infections. Care should be taken when introducing freshwater shrimps in order to improve a fishery as these carry an intermediate stage of the worm. Tapeworms are frequently also seen within the intestines of wild salmonids, notably *Eubothrium* (Plate 11), which may be present in large numbers without causing apparent harm to the fish or reducing its value for the table. In waters that contain large numbers of pike, plerocercoid stages of *Triaenophorus* (Plate 12) tapeworms may be found in every salmonid, and occasionally heavy infections result in thin and wasted fish.

Occasionally broodfish taken from the wild for spawning artificially show a collapse syndrome after spawning if they have been subjected to poor conditions or nutrition earlier in the cycle (Fig. 12.5). Affected fish are swollen and dropsical, and up to half may die.

Chapter 13
Prevention and Treatment of Disease

13.1 DISEASE PREVENTION

It has been emphasized that careful attention to good husbandry and fish management is essential in the control of disease. When disease strikes the land animal farmer, he or she is often tempted to assume it is somebody else's fault (e.g. the supplier of juvenile replacement stock or the feed supplier). This may or may not be true, but fish farms have the added risk of disease entering via the water supply. The use of borehole water or gravel filters on e.g. hatchery water supply greatly reduces the likelihood of many diseases entering the hatchery, especially parasitic infections. Others, notably virus diseases, must be prevented from entering the farm lest they become established. Such diseases commonly enter a farm along with eggs or live fish. They can also enter via the water supply, wild fish, birds and their droppings, fish food, and on the hands and especially the boots of farm personnel and lorry drivers. So the top priority must be to organize the farm in such a way as to prevent disease entering in the first place, but such preventive medicine is also about systematic hygiene and husbandry in order to minimize stress during day-to-day operations.

The first line of defence is national disease control legislation to prohibit the importation into the country of fish disease organisms, especially certain virus diseases. Obviously salmonid eggs should only be ordered from a farm that has been certified in writing by a competent laboratory as being free of the important virus diseases and BKD. Before arrival at their destination, the eggs should be disinfected, transferred into a clean utensil, and the original packaging burnt. An iodophor disinfectant should be used (e.g. 1% w/v diluted 1:100) into which the eggs should be dipped for 10 minutes. Such iodophors are very acidic and must thus be neutralized with a suitable buffer (e.g. sodium bicarbonate) to prevent harming the eggs at disinfection.

Live fish should never be introduced from another source, unless absolutely necessary. Suppliers of trout fingerlings or salmon smolts must be required to offer a detailed health history, with certified freedom from the most important diseases, particularly viral diseases; if in doubt specialist veterinary advice should be sought and the fish in question examined on the supplier's premises before the

purchase is confirmed. The track record of suppliers and health performance of their stock at on-growing sites is also important. In
addition, it is now normal to insist that the fingerling or smolt supplier arranges for vaccination against certain infectious diseases to be
undertaken at the hatchery prior to the fish being transferred to the
purchaser's farm for on-growing (see 13.4 'Vaccination'). When an
on-growing site is stocked with replacement juvenile fish, this should
be undertaken so as to keep different stocks of fish separate in order
to minimize the chance of disease transmission between them. For
example, trout fingerlings should be placed in special quarantine
ponds, where they can be held completely separate from other fish
for at least a month. The situation with seawater farms is more
complicated and the preferred solution for salmon on-growing farms
is to use different sites for different generations, i.e. to stock smolts
well away from the previous year's smolt generation. This allows
seawater sites to be farmed on a rotational basis, preferably with a
fallow period for the sites to recover at the end of each production
cycle.

Care should therefore be taken not only to avoid disease entering
a farm, but also to ensure that if a disease appears in one tank, pond
or cage unit, cross-infection to other rearing units will not necessarily
follow. In order to minimize stress, good water quality must be
maintained and good husbandry practice followed, with proper attention to stocking density, grading, tank or net maintenance, as well
as adequate feeding of good quality diets (see Chapter 10). It is
preferable to permit only the supervisor to enter the hatchery building and he or she should be required to walk through a footbath of
disinfectant solution on entry. If possible, there should be a different
dip-net and cleaning equipment provided for, and restricted to, each
rearing unit. Transfer of fish between different rearing units should
be kept to the minimum necessary. Any dead fish should be removed
daily, if possible, as they represent an indicator of health problems as
well as posing a contamination threat to the rest of the stock. Dead
fish should be disposed of by incineration or burial in a lime pit or
landfill well away from the farm site. Birds and rodents should be
discouraged by use of predator netting, scarers, trapping, etc. For
purposes of hygiene, the iodophor disinfectants are preferable because of their virucidal activity. They are extremely toxic to fish and
will precipitate a fish kill if they enter the water supply. The presence
of any organic matter markedly reduces the effectiveness of
iodophors, and equipment should be cleaned, e.g. with a pressure
hose, before disinfection.

Following harvest, cage units should be left fallow for 1–3 months
or longer before restocking, and earth ponds should be drained completely and allowed to remain dry for a month each year; sunlight has
a marked disinfectant action on ponds (and nets) and this is therefore
best done in the summer. If a stronger action is required, e.g. following an outbreak of whirling disease, quicklime (calcium oxide) may

be spread over the bottom of the pond at an application rate of $0.5\,\text{kg/m}^2$. Following an outbreak of virus disease on a freshwater pond farm, it may be desirable to kill out the entire farm, disinfect and then restock. In this case, earth ponds should be drained before spraying with a mixed solution of Teepol and sodium hydroxide (caustic soda). The mixture comprises 1 part of Teepol to 25 parts of 0.1 Normal sodium hydroxide, and it is sprayed at a rate of $3\,\text{l/m}^2$. If concrete or fibreglass tanks are being cleared of virus infection, then iodophor solutions are best used after scrubbing. In this situation it should always be borne in mind that if wild fish have contracted the infection, they are likely to remain carriers and may subsequently act as a source of reinfection.

13.2   RECORD KEEPING

This is an appropriate stage in the text to emphasize the crucial importance of maintaining good records. Initial site evaluation should provide details of seasonal temperature profile and water chemistry; for freshwater farms, factors such as water hardness and drought flow rate determine overall feasibility. However, systematic regular recording of various factors in the farm diary allows more efficient management as well as the opportunity to monitor trends in stock health and husbandry performance. These factors can be summarized as follows:

- Stock origin and numbers of fish in each rearing unit.
- Growth data and daily feeding rates for each rearing unit.
- Mortality data for each rearing unit.
- Details of any treatments given and drug withdrawal periods.
- Water temperature and water quality factors (e.g. dissolved oxygen pH, ammonia – especially for freshwater farms).

In the case of freshwater hatcheries or farms relying on running water, flow rates are critical; water flow is most commonly measured simply by determining the time taken to fill a container (or pond) of known volume. Accurate knowledge of the usable volume of pond space available, together with the flow rate data, should allow estimation of the stocking rates to be used. Water temperature data is also required and permits estimation of feed requirements for a certain stock of fish of given unit weight. Particularly under very intensive conditions at high water temperatures, monitoring of water chemistry factors, such as dissolved oxygen and total ammonia, helps to show when danger levels are being approached. pH recordings will allow calculation of free ammonia present from the value of total ammonia, and under marine conditions a salinometer will show any fluctuations in salt concentration. Losses should be counted daily and, when incorporated into the feed consumption and conversion efficiency calculations for a given rearing unit, often give valuable evidence of underlying disease problems.

Any treatments undertaken should always be amply documented, particularly the concentrations of chemicals used and all the calculations involved, the duration of treatment, the date and time of day, the sequel, etc. Such records are necessary in order to compile a unique treatment schedule to advise the correct dosages for different conditions based upon experience in the particular farm under study. In order to be able to comply with statutory drug withdrawal periods (see 13.3 'Control of medicinal products') and to demonstrate compliance if the farm is inspected, it is particularly important to record full details of any medicine being used. If the medicine in question has a 'degree day' formula for calculating drug withdrawal, then water temperature has to be monitored daily from the day the treatment ends. The use of unlicensed products under veterinary supervision or of licensed products outwith their datasheet recommendations, will normally require a standard withdrawal period (500 degree days in the UK, e.g. 50 days at 10°C). This also means recording the identity of rearing units and the numbers of fish to which the medicine is being administered, so that it is quite clear when the withdrawal period comes to an end and the fish can be legally offered for sale.

Appendix 1 gives a list of useful conversion factors.

13.3   CONTROL OF MEDICINAL PRODUCTS

The use of veterinary medicinal products, including their use in aquaculture, is now becoming more controlled internationally. For example, European Union (EU) legislation insists that before a product can be approved for commercial use, the safety, quality and efficacy of the product must be properly demonstrated. The safety aspect covers not only safety to the target fish species, but also safety to the operator (i.e. fish farmer, feed mill operator), safety to the environment (i.e. avoiding pollution or damage to the ecosystem), and safety to the consumer. Not only must the fish be wholesome to eat following treatment, but also any drug residues must be below the maximimum residue limit (MRL) set by the EU for the product in question. Before the EU grants a marketing authorization for any product, a withdrawal period has also to be set; this appears on the label and must be applied strictly following each treatment before the fish are harvested (or before they are stocked into a sport fishery).

The strict application of these control measures means that many products which have been used for many years by fish farmers are now becoming unavailable due to perceived hazards or simply due to the cost of attaining approval compared with the market size. Many drugs are prescription-only (e.g. antibiotic drugs for fish) and a veterinarian needs to authorize their use on each occasion. If there

Table 13.1 Medicated food calculation. How much of each drug to add to 1000 kg of feed (1 t) (dosage level = kg/t of feed)

Feeding rate %	Oxytetracycline pure (at 7.5 g/100 kg fish daily)	Furazolidone pure (at 7.5 g/100 kg fish daily)	Amoxycillin pure (at 8 g/100 kg fish daily)	Co-trimazine 40% active (sulphadiazine and trimethoprim) (at 3 g/100 kg fish daily)	Oxolinic acid pure (at 1 g/100 kg fish daily)
0.5	15.0	15.0	16.0	15.0	2.00
1.0	7.5	7.5	8.0	7.5	1.00
1.5	5.0	5.0	6.0	5.0	0.75
2.0	3.8	3.8	4.0	3.8	0.50
2.5	3.0	3.0	3.5	3.0	0.44
3.0	2.5	2.5	3.0	2.5	0.38
3.5	2.2	2.2	2.5	2.2	0.29
4.0	1.9	1.9	2.0	1.9	0.25
4.5	1.7	1.7	1.9	1.7	0.23
5.0	1.5	1.5	1.7	1.5	0.21
5.5	1.4	1.4	1.5	1.4	0.19
6.0	1.3	1.3	1.3	1.3	0.17

is no suitable drug approved for fish in a particular case, the veterinarian has the right within limits to prescribe the use of a drug approved for use in other animals. In such cases it is necessary to follow a standard withdrawal period of 500 degree days following treatment.

Even to use simple chemicals, such as formalin, which the fish farmer can buy over-the-counter without a prescription, it is normally necessary to have the permission of the authorities to release the product into freshwater or seawater (e.g. by a discharge consent). Although such chemicals have not previously been governed by medicines legislation, withdrawal periods are now being imposed in certain countries, and use of some chemicals (e.g. malachite green) may be banned altogether.

13.4   VACCINATION

Just as the use of vaccines has revolutionized human and land-animal medicine, fish vaccination has now become a cornerstone of modern fish husbandry. Not only do vaccines offer the possibility of being able to prevent diseases for which there is no effective treatment (e.g. virus diseases), but also the overall approach to fish diseases can be increasingly changed from curative to preventive. This in turn means that the fish farmer need not be over-reliant on current treatment methods, which are often short-term, only partly effective and require withdrawal periods to avoid the risk of drug residues. Vaccines stimulate the fish to produce an immune response to a particular disease-causing organism, which means the fish is protected from disease if it subsequently comes into contact with the organism in question. Such vaccines often comprised live but non-toxic strains of the infectious microbe, but nowadays most vaccines are killed during manufacture and an increasing proportion are made by techniques of genetic engineering.

The following salmonid pathogens have been used to make either experimental or commercial vaccines:

- Bacteria: *Aeromonas salmonicida*; *Aeromonas hydrophila*; *Vibrio anguillarum*; *Vibrio salmonicida*; *Yersinia ruckeri*.
- Parasites: *Ichthyophthirius multifiliis*; *Lepeophtheirus salmonis*.
- Viruses: IHN; IPN; VHS.

Already vaccines against furunculosis, vibriosis and ERM have been approved by the authorities and are being widely used on a commercial basis. Indeed, furunculosis outbreaks on many salmon farms have now been eliminated by injecting all smolts with vaccine prior to seawater transfer. Injection into the abdominal cavity appears to be the most effective vaccination route, but it is a skilled and time-consuming operation and the fish need to be sedated; currently this is the only way of vaccinating smolts for furunculosis as the vaccine emulsion is irritant and would damage fish flesh. Accidental self-injection by the vaccinator is a particular hazard and can cause gangrene in the affected finger or even sudden anaphylactic shock, so this form of vaccination should only be done by experienced personnel under veterinary supervision, and immediate medical treatment should always be sought if an operator is pricked or scratched. Some injectable salmon vaccines now offer protection against two or three different diseases with the same shot, i.e. vibriosis (*V. anguillarum*) and Hitra or coldwater disease (*V. salmonicida*) as well as furunculosis (*A. salmonicida*).

Vaccination by dipping fish into a bath of vaccine solution is much simpler and the preferred way of vaccinating trout and salmon fingerlings for ERM. Fry less than 1 g in weight are not protected by vaccination and they should be 2.5 g or even 5 g when vaccinated to give effective and long-lasting protection. An alternative method is to spray fish with the vaccine solution, but the ideal future method will be oral vaccination given in the feed, which is so far only a promising experimental technique. Undoubtedly there are further advances on the way as regards both vaccine technique and efficacy as well as extending the range of disease for which vaccines will be available, e.g. viruses.

13.5   TREATMENT METHODS

Fish are usually treated in one of three ways:

(1) adding chemicals to the water;
(2) adding chemicals to the feed;
(3) adding chemicals directly to individual fish.

13.5.1   Adding chemicals to the water

For treatment purposes, soluble chemicals are commonly added to the water in one of four different ways: as a dip, flush, bath, or flowing treatment. Most of the chemicals used are themselves toxic to fish to some degree. Treatment always imposes stress on fish, which are already weakened anyway. In general therefore, treatment should only be undertaken after careful evaluation of the circumstances and must always be considered a necessary evil. A number of precautions must be taken whenever chemicals are added to the water for treatment purposes.

(1) Do not feed for 24 hours prior to the treatment.
(2) Use plastic utensils for mixing, and never use galvanized containers.
(3) Ensure that any calculations of dosages are based upon accurate water flows and usable volumes of rearing units. Have your arithmetic checked independently.

(4) Treat first thing in the morning or at minimal water temperatures.
(5) Always carry out an initial trial treatment with a few fish.
(6) Wait 12–24 hours after the trial treatment before carrying out the main treatment if the trial is successful.
(7) Watch the fish continuously during treatment and be ready to flush rapidly with freshwater, or use oxygen diffusers or aerators should the fish become distressed.
(8) Wherever possible monitor dissolved oxygen levels throughout the treatment.
(9) Only repeat the treatment if absolutely necessary and not within 30 hours of the first treatment.
(10) Keep a comprehensive written record of what is done and the results.

For dip treatment, the solution is made up in a container into which a net containing the fish is dipped for a few seconds. Flush treatment involves adding the chemical to the inlet of a pond so that it runs through the system as a flush. In ponds with a high water exchange rate, e.g. raceways, where exposure to the drug is shorter, doses need to be higher (and therefore more critical). In certain systems, e.g. earth ponds, it is sometimes necessary to adopt a compromise between a bath and a flush.

For bath treatment, the water flow through a pond is stopped and the fish are bathed in the solution for a certain period of time; great care is needed to avoid killing the fish due to oxygen lack. When treating fish in floating cages by means of a bath (e.g. in marine farms), the total enclosure method is used; this involves surrounding the fish with a tarpaulin or plastic sleeve and crowding them by raising the production net somewhat, before adding the chemical while at the same time vigorously oxygenating the water through diffusers. No matter what method is used to set the treatment tarpaulin and to crowd fish, it should be set so that it can be removed or cut free from the cage as quickly as possible if an emergency arises during treatment. When the tarpaulin is removed at the end of the treatment, the chemical is soon diluted and dispersed, particularly where there is tidal exchange.

A large number of ponds or hatchery tanks can be treated simultaneously and accurately by means of a constant volume delivery pump or siphon. In this flowing treatment method, the pump is simply attached to the farm's intake and a constant volume of chemical is injected into the water supply to give the required concentration over the appropriate treatment period.

13.5.2 Adding chemicals to the feed

Certain types of drug are incorporated into the fish food in order to treat systemic bacterial infections or internal parasites. The main drawback with medicated food is that it is only effective if it is eaten, whereas many acute bacterial diseases cause the fish to go off their food. Also there may be a considerable time lag between diagnosing the disease and receiving from the feed mill a specially made batch of feed containing the appropriate drug, although it is sometimes possible for the farmer to coat pellets on-site (e.g. by making up a solution or emulsion of the drug with water or fish oil before spraying it onto feed pellets using a cement mixer). But it is the problem of trying to ensure fish consume sufficient food to receive the correct dose of drug that can cause most difficulty, relying as it does on feeding rates. The latter are influenced by water temperature, size of fish, possibly also by inappetence due to the disease, and in some cases by reduced palatability of the food due to the presence of the drug itself. Finally, it is important to observe the statutory drug withdrawal period following the treatment period before the fish can be harvested, sold or released into angling waters (see 13.3 'Control of medicinal products').

13.5.3 Adding chemicals directly to individual fish

Vaccination of individual salmon smolts has now become routine by means of injection. This involves prior sedation of the fish before intraperitoneal injection using repeat multidose syringes operated by trained personnel to minimize stress. Treating individual fish is largely restricted to valuable broodstock which will not enter the food chain, for example, the use of HCG (human chorionic gonadotrophin) injections in spawnbound fish to facilitate stripping of their eggs and the use of malachite green for painting large patches of skin fungus after prior anaesthesia.

Despite the reduction in medicament usage due to vaccination, mass treatment of salmonid fish using drugs in the water or food is still a regular practice, which requires particular skills if it is to be done safely and effectively. This will therefore be discussed in more detail by reference to the specific therapeutic agents used.

13.5.4 Specific treatment – chemicals added to the water

Formalin
Formalin is extremely useful for the treatment of external parasitic infections of skin and gills, especially *Costia* and other protozoans, and also monogenetic flukes. Formalin is a 40% solution of formaldehyde, and care must be taken to ensure that it is not contaminated with paraformaldehyde, which forms a white precipitate at the bottom of the bottle and which is very toxic to fish.

Formalin has an irritant action on the respiratory membranes of humans as well as fish. It must therefore be handled with caution for personal safety and used with extreme care as a treatment. Where possible, microscopic evidence of parasitism should be obtained before treatment is undertaken and gills of treated fish should be examined before any further treatment is contemplated.

Formalin removes oxygen from solution and this effect reaches a peak approximately 24 hours after it has been added to a pond. If formalin cannot be completely eliminated from the water in a fish pond soon after treatment, then care should be taken to prevent oxygen deficiency occurring, e.g. by using aerators.

When used as a bath, formalin should be thoroughly mixed into solution in order to achieve a final concentration of 1:5000 (200 ppm). Where legally permitted, one may   add a few drops of malachite green to the formalin initially, as the green dye will assist observation of adequate mixing throughout the pond. Under conditions of high water temperature, it may be advisable to use a concentration of 1:6000 (167 ppm) and the duration of treatment should never exceed 1 hour. This is a common way of treating fish in raceways during which it is always advisable to bubble air or oxygen into the system continuously.

Because of the slow water exchange rate involved for treating fish in earth ponds, it is usually necessary to lower the water level and adopt a compromise between a bath and a flush. The pond is lowered to half its normal depth and the quantity of formalin calculated which will ensure a final concentration of 1:5000 (200 ppm) with the pond half full. The required amount of formalin is mixed with water in a large plastic bucket and is slowly siphoned into the inlet monk over a period of 20 minutes. Throughout this period the water continues to flow through the pond, and when all the formalin has been added the pond is filled up again.

Benzalkonium chloride

Benzalkonium chloride is a blend of quaternary ammonium compounds which is of particular value in treating bacterial gill disease among fry and fingerlings. Toxicity of benzalkonium chloride to fish is markedly dependent upon the hardness of the water. In fry tanks the concentrations given below may be used as a bath.

Water hardness (ppm of $CaCO_3$)	Concentration of bath
Less than 100	1:1 000 000 (1 ppm)
100–200	1:500 000 (2 ppm)
More than 200	1:250 000 (4 ppm)

The chemical should be added slowly over the entire surface of the tank, and treatment should be for 1 hour, or less if the fish become in any way distressed.

For use in earth ponds, it is necessary to lower the pond to half its normal depth (compare formalin). The quantity of benzalkonium

Plate 51 (a) Abraded snout and eye of fish stung by jellyfish.

(b) Pale and swollen gills of Atlantic salmon with severe jellyfish stings to the branchial tissues. (Courtesy of Ian MacArthur and Captain Daniel Paterson.)

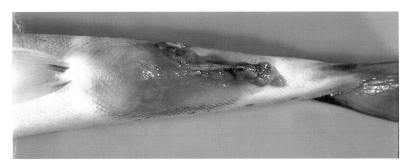

Plate 52 Autumn aeromonad disease. The lesions of this disease are bright-red in colour and often found near the vent. An added complication in this fish is the prolapse of the anal and urogenital area.

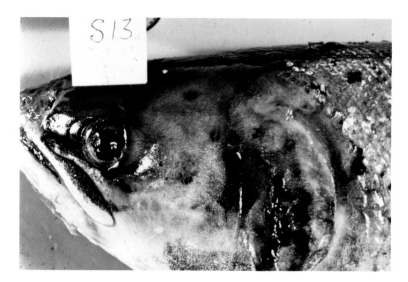

Plate 53 Uncomplicated UDN in salmonids. (a) The earliest stage of UDN, seen in Atlantic salmon as they enter freshwater. The very small ulcer with the greyish halo around it may heal or develop to the second stage where it becomes infected by waterborne bacteria or *Saprolegnia* fungus.

(b) A later stage in the early UDN lesions on fresh run Atlantic salmon. There are several ulcers on the side of the head and usually such lesions are symmetrically placed on either side.

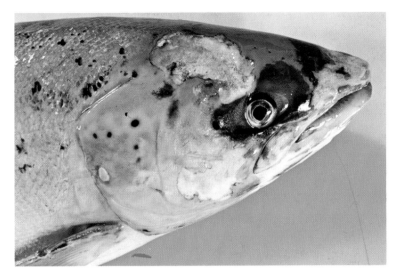

(c) A healing UDN lesion. This fish was kept in freshwater with added malachite green fungistat. This prevented the secondary infection and the picture shows the whitish scar tissue completing the cover of the raw ulcer.

(d) Severe UDN in a large sea trout. This fish had been in the river for some time, but although the lesion was long-standing, it had not become secondarily infected.

Plate 54 Skin cancer in an Atlantic salmon. This malignant tumour is affecting both the upper and lower jaw. (Courtesy of G. Macgregor.)

Plate 55 Cystic spleen of wild brown trout. (Courtesy of G. Macgregor.)

Plate 56 Cystic peritoneal anomaly in Atlantic salmon. (Courtesy of G. Macgregor.)

Plate 57 *Acanthocephalus* infection. The parasites are attached to the lining of the gut by the thorny proboscis at the head. The intestine is reddened and the gut mucus may be haemorrhagic.

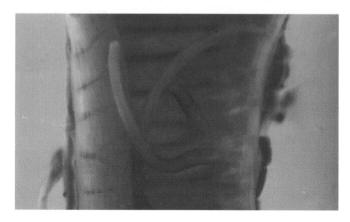

chloride is calculated which will ensure a concentration of 1 : 1 000 000 (1 ppm) with the pond half full. This amount is then mixed in a plastic bucket and, to ensure careful mixing, one-tenth of the total is mixed with a whole bucketful of water and poured into the inlet monk, and the process is repeated. After all ten bucketfuls of the drug have been used and the full dose administered, the pond is allowed to fill up again.

Chloramine-T

Chloramine-T is valuable for the treatment of external parasite infestation and as a skin disinfectant in fry and fingerlings. Toxicity of chloramine-T to fish is markedly dependent upon the hardness of the water (compare with benzalkonium chloride). In soft water with a low pH, a concentration of 2.5 ppm may be used as a bath for 1 hour. In hard water with an alkaline pH, trout farmers frequently use a concentration of 8.5 ppm for 1 hour. It is also used for salmon in seawater to treat surface wounds, in which case a concentration of up to 20 ppm may be used for 1 hour.

Malachite green

Malachite green is very effective in treating fungus infections, particularly *Saprolegnia*. It is also useful in treating PKD and white spot disease (Ich). However, malachite green is suspected of being a carcinogen and it has therefore been banned from use for fish in some countries, including the UK, and should always be handled with care using protective clothing. It is obtained as a crystalline solid which dissolves in water to give a green dye. It is important at purchase to stipulate that zinc-free malachite green is required in order to avoid lethal zinc toxicity problems (for the same reason, galvanized containers must always be avoided).

As a dip treatment, malachite green may be used at a concentration of 1 : 15 000 (67 ppm) for a period not exceeding 30 seconds. As a bath treatment, it may be used at a concentration of 1 : 500 000 (2 ppm) for 1 hour for fry and fingerlings, and 1 : 1 000 000 (1 ppm) for 1 hour for adult fish. For flush treatment in Danish-type earth ponds, a total dosage of 1 : 200 000 (5 ppm) is recommended. The required amount of malachite green is calculated and then is best added to the inlet monk over a period of approximately 2 hours (e.g. 5–6 tablespoonfuls at the rate of half a spoonful every 10 minutes).

Malachite green is highly valued as an antifungal agent for egg treatment. In this case a concentration of about 1 : 500 000 (2 ppm) for 1 hour is preferred. Since the water flow cannot usually be stopped, this requires a continuous drip of the chemical which may be arranged using a constant flow siphon or similar device.

Formalin and malachite green is an effective combination for controlling stubborn outbreaks of white spot disease (Ich). American workers have obtained good results using 3.68 g of zinc-free malachite green per litre of formalin. This stock solution is then applied to yield a treatment concentration of 1:40000 (25 ppm) by either of the methods proposed for formalin alone.

Dichlorvos

Dichlorvos is an organophosphorous compound widely used for the treatment of sea lice (*Lepeophtheirus, Caligus*). For bath treatment of lice-infested fish in cages, the total enclosure method is best after an initial 24-hour starvation period. After setting the tarpaulin and crowding the fish, dichlorvos is added at a concentration of 1 ppm active ingredient for up to 60 minutes, while oxygenating the water through diffusers. The product is dangerous to fish and fish farmers, so should be handled with care following codes of safe practice. Treatment should be stopped if fish become stressed and start burrowing into the net or rolling around.

Dichlorvos is only active against pre-adult or adult stages of the parasite and it is therefore often necessary to undertake a second treatment in order to kill the young parasitic stages after they grow through to adults. For this reason and because of the risk of lice developing resistance to dichlorvos, it is recommended that sea lice populations on salmon are regularly monitored. A drug withdrawal period of 4 days is needed after treatment before the fish are harvested.

Newer bath treatments for sea lice

There is concern about suggested safety hazards which may be associated with using dichlorvos and other organophosphorous compounds (e.g. azamethaphos). Newer alternative bath treatments have therefore now become available of which the most widely used is hydrogen peroxide. This has the advantage of releasing oxygen on contact with organic matter and thus it physically destroys lice (although large female lice often survive and may re-attach). Treatment time using total enclosure with tarpaulins is only 20 minutes and since hydrogen peroxide leaves no residues, there is a zero withdrawal period. It may be necessary to increase the recommended dose carefully at low water temperatures.

The synthetic pyrethroid cypermethrin appears highly effective against both mobile and attached stages of sea lice when used at an active concentration of 5 ppb as a bath for 1 hour. Until a withdrawal period is agreed by the authorities, the standard withdrawal period of 500 degree days should be followed.

Anaesthetics

Anaesthesia is used for a wide variety of handling operations with salmonid fish, ranging from vaccination and tagging of smolts to sedation of broodfish prior to stripping. Many different anaesthetics have been recommended for use with fish. While injectable anaesthetics are feasible and work well, the most widely used are all applied as drugs in solution which are ventilated (inhaled) by the fish over the gills. From there it is a short distance to the central nervous system, which receives its blood supply directly from the gills. It is analogous to gaseous anaesthesia of higher animals and humans. The inhaled drugs are also exhaled via the gills and the urine, so once the fish are transferred to clean water, excretion of the drug quickly allows recovery.

Usually the fish are simply immersed in a suitable concentration of the relevant drug. Unfortunately, maintenance of level anaesthesia is difficult, as is estimation of precise doses for the specific fish being anaesthetized, but provided the period of anaesthesia required is short, and the fish are rapidly returned to clean water, a broad-spectrum of dose regimes seems to be acceptable. Large fish with high fat levels in their tissues will take longer to anaesthetize and also much longer to recover, since they absorb considerable levels of anaesthetic within their fat and only release it gradually.

Three anaesthetics are commonly used with fish.

Quinaldine
Quinaldine sulphate, commonly known as quinaldine, is an oily liquid, more potent at higher pH levels. It is very effective, but because it is irritant to mucoid membranes, it can damage the cornea of the eye. It is relatively inexpensive and widely used by fishery biologists to capture fish in the wild.

MS222 (methane tricaine sulphonate)
MS222 is a very well-tried anaesthetic which has the advantage of being very soluble in water. It is normally used at 20–25 mg/l of water, though more may be necessary at lower temperatures. MS222 has a high anaesthesia/mortality ratio so is remarkably safe, but is irritant, and also it is an expensive drug compared with its alternatives.

Benzocaine
Benzocaine is widely used since it is both cheap and effective. Its only limitation is its insolubility. For this reason, it is normally prepared as a stock solution of 100 g/l dissolved in acetone from which small volumes are withdrawn and used for anaesthesia. It is normally active at levels of 30–40 mg/l of water and gives good rapid anaesthesia from which recovery is also rapid. It is less irritant than MS222 and is safer to humans and animals. If the stock bottle of acetone suspension is kept cool and dark, it is likely to last for at least a year without losing activity.

Miscellaneous chemicals added to the water

Trichlorphon
Trichlorphon is an organophosphorous compound which is closely related to dichlorvos. Trichlorphon formulations have been widely

used for treating copepod parasites, such as fish louse (*Argulus*) and anchor worm (*Lernaea*); it will also kill some monogenetic flukes, leeches and protozoa. Trichlorphon is sprayed on large ponds or added to the inflow to achieve a total concentration of 1:4000000 (0.25 ppm) and allowed to dissipate out of the system. Two applications at 1 week's interval are usually sufficient for large ectoparasites. The availability of trichlorphon for aquaculture use is becoming increasingly limited by regulatory concerns over the safety of organophosphorous compounds.

Potassium permanganate

Under conditions of oxygen deficiency, potassium permanganate ($KMnO_4$) can be of value in reducing BOD of the water (e.g. after formalin treatment of a heavily stocked pond in mid-summer). Addition of potassium permanganate to give a concentration in the pond of 1:1000000–1:500000 (1–2 ppm) is sometimes a useful emergency oxygenation measure if there is oxygen depletion or during installation of equipment such as aerators and pumps.

Sodium chloride

Iodine-free salt (sodium chloride) has been used for many years by salmon and trout hatcheries in treating fungal infections and external parasitism, e.g. costiasis (*Ichthyobodo*). It is most practical for treating alevins and fry at 0.5–2% (5000–20000 ppm) for up to 30 minutes. Larger trout may be treated for up to 30 minutes with a 3% bath (30000 ppm), but this should be stopped if fish start to look stressed.

Others

Various other chemicals (e.g. copper sulphate and methylene blue) are used on certain trout and salmon farms, but are not described here as it is considered that they are probably inferior in effectiveness and safety to those already considered.

Specimen calculations

Formalin treatment of trout for external parasitism in raceways

(1) Recommended dose is 1:5000 (200 ppm) formalin for 1 hour.
(2) Ascertain treatment volume of raceway, e.g. $30 \times 2 \times 0.5$ m deep = $30 \, m^3$ = 30000 litres.
(3) Calculate volume of formalin to give required concentration,

$$\text{i.e.} \quad \frac{30000}{5000} = 6 \text{ litres.}$$

Therefore in this example 6 litres (1.32 gallons) of formalin have to be carefully spread and thoroughly mixed over the entire length of the raceway to give a 1:5000 concentration at a depth of 0.5 m for 1 hour with appropriate aeration facilities.

Benzalkonium chloride treatment of trout in earth ponds

(1) Ascertain hardness of water, e.g. 150 ppm $CaCO_3$.
(2) Recommended dose for bacterial gill disease is 1:500 000 (2 ppm) benzalkonium chloride for 1 hour for this hardness.
(3) Calculate volume of pond when half full only, e.g. $30 \times 10 \times 0.5$ m deep = 150 m^3 = 150 000 000 ml.
(4) Calculate volume of benzalkonium chloride to give required concentration

$$\text{i.e.} \quad \frac{150\,000\,000 \times 2}{1\,000\,000} = 300\,\text{ml.}$$

Therefore in this example 300 ml (0.066 gallons) of benzalkonium chloride are poured into the inlet monk diluted in water to give 10 bucketfuls of solution. The water level is then raised to the normal height once again.

13.5.5 Specific treatment – chemicals added to the feed

Anthelmintics for treating tapeworm infestation

The preferred treatment for intestinal tapeworm infestation is either praziquantel or fenbendazole. Praziquantel is licensed for use in feed in Norway with a dose of active ingredient of 500 mg/100 kg fish/day for 1–2 days and a withdrawal period of 14 days. Fenbendazole is used at an active ingredient single dose of 500–800 mg/100 kg fish which is usually repeated between 2 and 4 days later; although in Norway the withdrawal period following this second dose is 30 days, in the UK a standard withdrawal period of 500 degree days is required. Although these drugs are effective against *Eubothrium* and other tapeworms within the gut, it must be borne in mind that they are no use in treating *Diphyllobothrium* plerocercoids outside the gut.

Drugs for treating sea lice infestation

Three in-feed drugs currently being tested show promising results in sea lice control and are likely to be approved for commercial use in due course. Ivermectin is a well-known anthelmintic for farm animals, now being prescribed by some veterinarians for use in salmon feed at an active ingredient dose of 2.5 mg/100 kg fish on a twice-weekly basis. To avoid any risk of ivermectin residues entering the food chain, farmers must observe the standard withdrawal period and ivermectin must only be given to salmon during their first sea year following transfer from the hatchery as smolts. In Norway, test licences have been issued for the use of in-feed diflubenzuron (0.6 g/kg feed for 14 days) and a related in-feed compound teflubenzuron (2 g/kg feed for 7 days) for treating salmon for sea lice during their first sea year only.

Antibacterial treatments

Most antibacterial compounds fall into one of three drug groups – sulphonamides, nitrofurans and antibiotics, – all three of which are used to treat bacterial diseases in fish. Sulphonamides were first used successfully against furunculosis in the 1950s, but used alone they are unpalatable to fish and at high doses can cause kidney damage. Nowadays a conventional sulphonamide, such as sulphadiazine, is formulated for fish in combination with a 'potentiator' such as trimethoprim; the resulting potentiated sulphonamide offers greater efficacy with fewer side-effects. Nitrofurans have been widely used for controlling fish diseases, but are now falling into disrepute due to consumer safety fears, although furazolidone is still widely used in some countries for bacterial and internal parasitic disease (*Hexamita*: see below).

Probably the most widely used broad-spectrum antibiotic for fish is still oxytetracycline. Quinolones, such as oxolinic acid and flumequine, have not realized their initial potential due to questions about drug resistance and long-term safety. Amoxycillin is a synthetic penicillin which is useful for treating salmon with furunculosis, and has also been used for treating rickettsial infection of salmon in Chile and rainbow trout fry syndrome due to *Cytophaga psychrophila* (or *Flavobacter*?) infection. Amoxycillin is cleared rapidly from fish tissues which offers the benefit of a short withdrawal period, but sometimes also means a greater risk of reinfection.

Many of these drugs are used in human medicine and because of the danger associated with drug resistance and the remote risk that such resistance could be transferred from fish to humans, it is important that they should not be used indiscriminately. Certain drugs of particular value in treating human disease, notably chloramphenicol, should never be used in fish destined for human consumption; the related compound florphenicol, however, has considerable promise for use in fish. In order to exert responsible control and prevent abuse, most countries insist that the antibiotics, sulphonamides and nitrofurans can only be obtained on veterinary prescription (see 13.3 'Control of medicinal products'). It is important that the entire course of treatment prescribed should be followed closely. Drugs must be administered only to the particular stock of fish for which they were intended and withdrawal periods observed scrupulously.

For acute bacterial infections in salmonids, the choice of feed additive should reflect which drug the particular organism is most sensitive to as well as the economics of treatment. This means that, wherever possible, the choice of drug should await the outcome of bacterial isolation and antibiotic sensitivity testing in the laboratory, as described in Appendix B. Table 13.1 should be consulted to enable rapid calculation of the amount of drug to be mixed in with a particular proportion of feed. It is important to remember that the use of feed additive drugs cannot reliably eliminate drug-sensitive pathogens from fish under treatment since it is impossible to guarantee achieving the tissue levels of drug necessary to abolish bacterial 'carriers' in the stock.

Oxytetracycline

Oxytetracycline is probably still the most widely used and cheapest antibiotic for treating acute septicaemias. The recommended dosage of active ingredient is 7.5 g/100 kg fish/day for 5–10 days. The drug withdrawal period depends on the formulation and country, but is generally in the region of 400–600 degree days.

Amoxycillin

Amoxycillin is used mainly for treating furunculosis in salmon. The recommended dosage is 4–8 g/100 kg of fish per day for 10 days. Since amoxycillin is cleared rapidly from fish tissues, a withdrawal period of only 50 degree days is needed, allowing salmon to be treated shortly before harvesting.

Potentiated sulphonamides

The combination of sulphadiazine and trimethoprim has a wide range of activity against most common bacterial septicaemias. The recommended dosage of combined ingredients is 3 g/100 kg fish/day for 5–7 days. The sulphonamide content is fairly unpalatable which can cause problems, especially at low water temperatures. The drug withdrawal periods are 350 degree days for Atlantic salmon and 500 degree days for rainbow trout.

Quinolones

Oxolinic acid has been widely used for treating acute bacterial infections (e.g. furunculosis), despite increasing concern about drug resistance and safety. The recommended dosage in freshwater is 1 g/100 kg fish/day for 10 days, but veterinarians prefer to increase this to 3 g/100 kg fish/day for effective use in seawater; the withdrawal period is 500 degree days. Flumequine is a related quinolone which is also used in some countries with the same freshwater dose rate as oxolinic acid, whereas the recommended dose of flumequine in seawater is 1.25–2 g/100 kg fish/day for 5–8 days. Newer quinolones are currently under evaluation (e.g. sarafloxacin, enrofloxacin).

Furazolidone

In addition to its antibacterial activity, furazolidone has been regarded as the most effective available treatment for *Hexamita* gut parasitism. The recommended dosage of active ingredient is 11 g/100 kg fish/day for 5–10 days. A major drawback has always been poor palatability, but furazolidone is now also regarded as unsafe in the human food chain and is therefore banned in many countries.

13.5.6   Calculating in-feed drug addition rates

The most critical calculation is the feeding rate. When you incorporate the chosen medicament into feed at the farm or at the feed mill, the feeding rate determines, at a given dose, how much drug goes into the diet. If you assume a feeding rate of 4% and the fish only eat 2%, they only get half the correct dose. So make sure you get it right! Be sure to feed all of the medicated feed first as you can always 'top up' with unmedicated feed afterwards.

Specimen calculation

(1) Decide the dosage and length of treatment. These are recommended in the text.
(2) Choose a feeding rate 0.5% lower than the one you have been using. Thus, a normal feeding rate of 3% means a medicated diet feeding rate of 2.5%.
(3) Find out the total weight of fish requiring treatment.
(4) Work out how much food you will need for the length of treatment, e.g. 400 kg of fish at normal feeding rate of 3% for 10 days:

$$\text{food required} = 400 \times \frac{2.5}{100} \times 10 = 100 \text{ kg}.$$

(5) Turn to Table 13.1 which will tell you how much drug to add to 100 kg of feed for the particular dosage and feeding rate. For example, oxytetracycline = 3.0 kg/t of feed at 2.5% feed rate. However, you need 100 kg of feed, and therefore need to add:

$$3.0 \times \frac{100}{1000} = 0.30 \text{ kg} = 300 \text{ g oxytetracycline (active)}.$$

Appendix A

THE MICROSCOPE ON THE FISH FARM

A microscope is a very expensive precision instrument which can only be used correctly if the basic structure and function is understood. Microscopes should always be covered when not in use, either in a case or by a waterproof, dustproof cover. Figure A1 shows the parts of a standard microscope with a built-in light source. Many older microscopes have a mirror for purposes of illumination.

Microscopes are available in many price ranges. You get what you pay for, and good lenses are very expensive. A new microscope with built-in light source, suitable for fish farm work, cannot be bought for less than £2000 (approx $3000). Good second-hand microscopes are sometimes available from medical students and older brass ones with good lenses can be excellent, but inexpensive modern microscopes are not to be recommended. Eyepieces should be 10× and the objectives most used are 10× and 40×, giving total magnification of 100× and 400×, respectively. If bacteriological examinations are to be carried out then a 100× oil immersion lens is also necessary. Since a microscope on a fish farm is often used for examining wet mounts of material containing moving specimens, a moving stage, which can be readily manipulated, is highly recommended.

Use of the microscope

The stand should be placed on a firm surface with a light source, such as the sun, or a torch or electric bulb available if the microscope is not supplied with a light built in. The light should be switched on and the mirror focused as necessary.

The slide to be examined is placed between the clips on the viewing stage or between the arms of the moving stage. The appropriate objective lens is swung into position and the condenser (if fitted) focused by means of the condenser control knob, to allow good illumination. If bacteriological examination is to be carried out then a 100× objective allowing 1000× magnification has to be used with oil being interspersed between the lens and the specimen to allow crisp viewing.

Focusing

Turn the coarse adjustment knob, *whilst watching the slide and objective, not looking down the microscope,* until the objective is about 5 mm

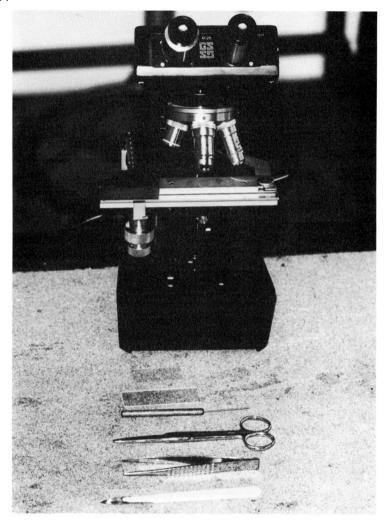

Fig. A1 Post-mortem equipment required on the fish farm. The microscope must be capable of magnification up to 400×, and built-in light and a moveable stage are of great advantage.

from the specimen, then look down the eyepiece and turn the coarse adjustment knob until the specimen comes roughly into focus. Then it should be brought into perfect focus with the fine adjustment. Because the 40× lens has a depth of focus only one-fifth that of the 10× objective, it is not possible to extract the maximum amount of information from a specimen using the 40× objective unless the fine focus control is constantly moved up and down through the planes of focus.

When stained bacterial smears are to be examined, a few drops of inert microscope oil must be placed in the centre of the smear. The slide is then focused under low power and the 100× lens swung across the light path. The oil immersion lens has to be focused very carefully with only very slight movement of the fine adjustment. Oil lenses are very expensive, and extensive racking down can result in them breaking the slide and damaging their finely polished surfaces.

Care of lenses

The lenses are the most expensive and fragile components of the microscope and should be treated with the utmost care. The oil immersion lens should always be cleaned after use, in order to remove the oil. This must be carried out using a special fine lens tissue, dipped in a little xylene. This method can also be used to clean low power lenses should they become dirty. Do not use methylated spirits or acetone to clean lenses as these may damage the lens cement.

Appendix B

LABORATORY DIAGNOSIS OF BACTERIAL INFECTION

A microscope is an essential piece of equipment on the well-managed fish farm in order to allow the proper diagnosis of parasitic and certain other common conditions by examining wet mount preparations (see Fig. 5.1).

The intensification of salmonid farming over the past few years has allowed a greater level of biological expertise to be available on the farm, and in some farms it is now considered appropriate for the hatchery biologist to carry out simple bacteriological diagnostic work. It must be stressed that proper and detailed bacteriological examination requires sophisticated equipment and the experience of a specialist veterinarian or professional microbiologist, but with some training the able hatchery biologist should be able to diagnose the more common problems and carry out at least the more straightforward, qualitative bacterial antibiotic sensitivity tests.

Equipment

Most of the pathogenic fish bacteria are relatively undemanding in their growth requirements and grow well at normal room temperatures (18–22°C). Thus, only a limited amount of equipment is required (Table A1). The main problem for fish farmers who wish to do their own bacteriology is obtaining regular supplies of fresh sterile dishes of nutrient media for inoculation. These can only be stored for a limited period, and so it is only worth considering doing on-farm bacterial culture if a reasonable throughput of material is anticipated.

Techniques

Wet mounts

Where mucus samples from the gill or skin are to be examined for the presence of Myxobacteria, as in the case of columnaris disease or bacterial gill disease, scrapings of material from the affected tissue are placed on a slide, with a drop of water and a coverslip. Examination under the 40× objective will reveal relatively long slender bacteria which can be seen moving sinuously around the field (Fig. A2).

Table A1 Equipment required for basic bacterial isolation and examination on the farm

---

Microscopical examination
- Good quality compound microscope with 100× oil immersion objective lens.
- Bunsen burner or spirit lamp for fixing smear preparations.
- Set of bacterial stains for Gram's method (see Appendix C).
- Wash bottle.
- Good quality microscope slides and diamond marking pen.

Cultural examination and sensitivity
- Inoculating loop.
- Primary culture agar plates (normally TSA).
- Antibiotic sensitivity discs.

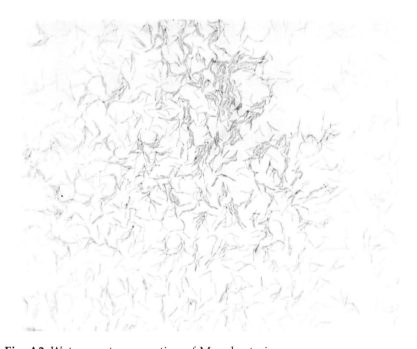

Fig. A2 Wet mount preparation of Myxobacteria.

Stained smears

When a lesion appears to be associated with a bacterial disease, it is often useful to prepare a stained smear of material from the affected area, and examine it under the highest power (oil immersion × 100) lens in order to determine which types of bacteria are present within the lesion. Equally, when bacteria have been isolated on an agar plate as appropriate, it is important to determine their features and staining reactions by removing a little from the plate to make a smear. The smear is prepared by placing a drop of water on a clean slide and emulsifying a small amount of the material from the tissue, or a small part of the colony in it. After the slide has been marked for identification, it is passed through a bunsen flame a couple of times to heat-fix the material to the slide, and then it is ready for staining to make the bacteria visible in the microscope.

Gram's stain not only stains the bacteria but also allows the different species of bacteria to be divided into two main categories, the Gram-positive bacteria, which stain blue-black with this staining method, and Gram-negative, which stain red. Most fish pathogens are Gram-negative, but important Gram-positives include *Renibacterium salmoninarum*. *Mycobacterium marinum* and related bacteria do not stain by the normal Gram's staining method because of the nature of their surface and special acid-fast staining techniques are required to demonstrate them. Plates 18 and 19 show typical Gram-negative and Gram-positive stained smears.

Isolation of bacteria from pathological material

Usually tissue suspected of containing bacterial pathogens is inoculated onto a solid medium or plate of agar gel containing proteins and other ingredients required by the bacteria to sustain growth. Even if aseptic technique is practised at sampling, it is possible to have more than one bacterium type growing on a plate, derived from a mixed infection in the lesion. Thus, it is important to ensure, when inoculating the plate, that growth of the bacteria occurs in such a way that individual colonies of bacteria (derived from growth of a single cell) develop in at least some parts of the plate. These can then be removed to sub-culture, as a pure culture, and can also be used for carrying out antibiotic sensitivity testing (Fig. A3).

When the bacterial growth is examined after 2 or 3 days at room temperature, a variety of raised shiny colony types will usually be

Fig. A3 Sampling fish for bacteriology. (a) The requirements for sampling. (b) Flaming the scalpel to sterilize it. (c and d) Searing the surface to be sampled and sterilizing the inoculating loop. (e) Lifting the edge of the lesion. (f) Resterilizing the loop. (g) Sampling from below the skin surface. (h and i) Inoculating the surface of the agar medium on the petri dish. (j) Opening the abdomen with a sterile scalpel. (k) Pushing the viscera aside with a resterilized scalpel. (l) Sampling with the sterile loop from the kidney, taking care not to touch other organs.

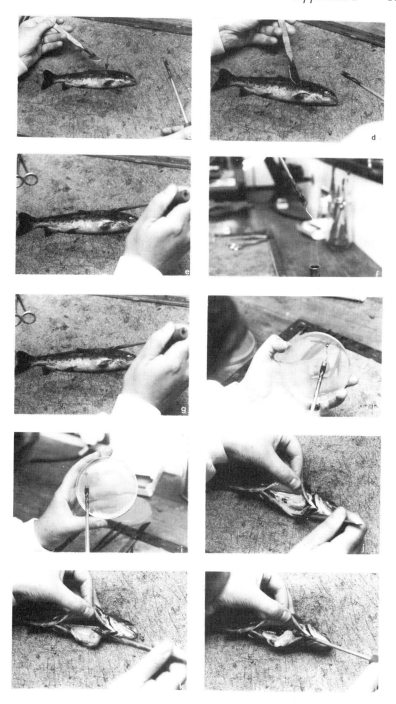

Fig. A3 *Continued*

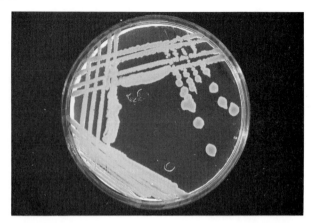

Fig. A4 Bacterial colonies growing on an agar plate. The individual colonies are on the right side of the plate.

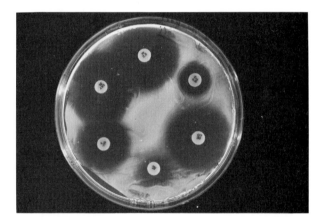

Fig. A5 Sensitivity discs on an inoculated plate showing zones of inhibition.

observed growing on the surface. If a bacterium is present in the tissues in any significant numbers it will usually predominate in the culture on the plate (Fig. A4). Small pieces of typical individual colonies should be used to sub-culture from these and also to prepare Gram smears for morphological examination of the bacterial structure.

Antibiotic sensitivity
When a dominant bacterial species is isolated and thought likely to be responsible for a particular disease, it is valuable to know what antibiotics the organism is sensitive to, and to which it is resistant, in order to help choose the appropriate drug therapy. For this purpose, a plate is liberally seeded with bacteria from a suspect colony by smearing material from the colony over the surface of the plate with

an inoculating loop. A commercially produced impregnated antibiotic sensitivity disc containing measured amounts of various antibiotics is then placed in the centre of the plate, which is incubated until good growth is obtained. It is then examined and the presence of active inhibition of growth is indicated by a zone around the appropriate antibiotic in which no colonial growth of bacteria occurs (Fig. A5).

It is important to remember that this technique is only able to give qualitative information and only applies to the laboratory situation. Although it provides a guide to what the effect of different antibiotics would be likely to be in the fish, it is not always completely reliable because of other factors, such as the effect of the fish's metabolism on the antibiotic when it is administered.

Appendix C

LABORATORY TECHNIQUES AND CHEMICALS FOR SALMONID DISEASE DIAGNOSIS

Staining bacteria by Gram's method

Reagents required

(1) Ammonium oxalate – crystal violet solution

Crystal violet (Colour Index No. 42555)	2 g
Ethanol (95%)	20 ml
Ammonium oxalate	0.5 g
Distilled water	80 ml

Dissolve crystal violet in ethanol. Dissolve ammonium oxalate in distilled water. Mix the two solutions, allow to stand for 24 hours and filter.

(2) Iodine solution

Iodine	1 g
Potassium iodide	2 g
Distilled water	300 ml

Dissolve potassium iodide and then iodine in about 5 ml of distilled water. Add remaining volume of water.

(3) Safranin solution

Safranin (Colour Index No. 50240)	0.25 g
Ethanol (95%)	10 ml
Distilled water	90 ml

Dissolve safranin in ethanol and add distilled water.

Method

(1) Apply ammonium oxalate – crystal violet solution to heat-fixed smear for 1 minute.
(2) Wash with water.
(3) Apply iodine solution for 1 minute.
(4) Tip off iodine solution.
(5) Decolourize with alcohol/acetone (95%/5%) until no more violet colour emanates from the smear.
(6) Wash thoroughly with water.
(7) Apply safranin solution for 2 minutes.

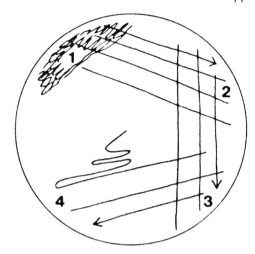

Fig. A6 Sequence of plate inoculation for bacterial culture.

(8) Wash with water, drain and/or blot dry and examine.
Gram-positive micro-organisms show as blue/purple.
Gram-negative micro-organisms show as pink/red.

Plate inoculation procedure for culture of bacteria

(1) Inoculate tissue sample on a segment of the surface of the culture medium (**1**).
(2) Flame inoculate loop until red hot, allow to cool and touch edge of uninoculated area of medium to ensure coolness.
(3) Spread part of sample over about one-quarter of the plate by making three or four parallel streaks with the loop (**2**).
(4) Repeat streaking procedure as shown (**3** and **4**), flaming and cooling the loop between each sequence.
(5) Label the underside (not the lid) of the plate and incubate.

For the purification of mixed cultures and sub-culture on solid media, follow the same procedure using a small portion of a bacterial colony for the initial inoculum (**1**).

Fixation

If fish tissue samples are being collected for submission to the laboratory for pathological diagnosis, they must be preserved in embalming or fixing fluid. It is very important that the samples are removed from the fish immediately after it has been killed. Tissue from dead fish will have deteriorated so that it is usually no longer of any value. Small blocks of tissue, no larger than 1 cm cubes, are placed in considerably greater (×40) volumes of fixative for at least 24 hours before submission to the laboratory.

The easiest fixative to make is 10% formalin, but a better one for fish is phosphate buffered formalin, made up as follows:

40% formaldehyde	100 ml
Tap water	900 ml
$NaH_2PO_4 H_2O$	4 g
$Na_2 HPO_4$	6 g

Appendix D

CONVERSION FACTORS

Weight
1 lb (16 oz)	= 454 g
1 oz	= 28.4 g
1 kg (1000 g)	= 2.2 lb (35.27 oz)
1 ton	= 2240 lb
1 metric tonne	= 2204 lb

Volume
1 gallon (8 pints or 4 quarts)	= 4.55 litres
1 litre (1000 cc or 1000 ml)	= 0.22 gallons
1 litre	= 1.76 pints (0.26 US gallons)
1 litre	= 35.2 fluid oz
1 cubic foot	= 28.3 litres
1 cubic foot	= 6.22 gallons
1 cubic yard	= 0.764 cubic metre
1 cubic metre	= 1.308 cubic yard
1 cubic metre	= 35.31 cubic foot

Length
1 inch	= 25.4 mm (2.54 cm)
1 foot	= 30.48 cm
1 yard	= 0.914 m
1 metre	= 39.37 inches

Area
1 sq yard	= 0.836 sq metres
1 sq metre	= 1.196 sq yard
1 acre	= 4140 sq yard
1 hectare	= 2.47 acres

Water equivalent
1 gallon	= 10 lb (4.54 kg)
1 cu ft	= 6.23 gall (28.3 kg)

Flow rate
1 gallon per minute (gpm)	= 75.7 ml per second = 75.7 cc per second
1 gallon per hour	= 1.26 ml per second
1 litre per second	= 13.2 gpm

| 1 cu ft per second | = 28.3 litre per second (373.8 gpm) |
| 1 cu ft per second | = 538 272 gallons per day |

Temperature

On the Celsius scale, 0°C and 100°C represent the freezing and boiling points, respectively of water (at standard pressure).

To convert (°C) degrees Celsius to Fahrenheit (°F), use the formula:

$$\left(°C \times \frac{9}{5}\right) + 32 = °F$$

ppm	= parts per million	
1 ppm	= 1 mg/litre	= 1:1 000 000
1 g/litre	= 1:1000	= 1000 ppm

Further Reading

BOOKS

Frerichs, G.N. & Millar, S.D. (1994) *Manual for the Isolation and Identification of Fish Bacterial Pathogens*. Pisces Press, Stirling.

Inglis, V., Roberts, R.J. & Bromage, N.R. (1993) *Bacterial Diseases of Fish*. Blackwell Science, Oxford.

Roberts, R.J. (1989) *Fish Pathology*. Baillière Tindall, London.

Willoughby, L.G. (1995) *Fungi and Fish Diseases*. Pisces Press, Stirling.

Wolf, K. (1988) *Fish Viruses and Fish Viral Diseases*. Cornell University Press, Ithaca, New York.

JOURNALS

Aquaculture. Elsevier Science, Amsterdam.

Aquaculture Research. Blackwell Science, Oxford.

Diseases of Aquatic Organisms. Paul Parey, Berlin.

Journal of Fish Diseases. Blackwell Science, Oxford.

Index